Transformations Used For Electrical Systems

Doug York

© 2017 Douglas York

Transformations

Transformations

To Francis X Bostick,
my friend, teacher and mentor.
Thanks for appreciating me for what I am.

Transformations

Transformations

This booklet is written to help with the understanding of mathematical relationships and transformations that are used in the analysis of three- (or more) phase power sources, motors, motor drives and the transmission systems between them. It is intended as an introduction for those new to the study of ac management, but can also provide a coordinated review for more experienced designers.

In studying the various transformations in the literature, one finds that the different authors have their own ways of describing their favorites. This booklet first sets up a set of tools with which to consistently describe many of these relationships, then shows their similarities and differences, and above all highlights their simplicity. We will work primarily with three-phase systems in our development, although some of the techniques are usable with higher order polyphase systems.

This book is short, because there is not much repetition – it is written to be re-read, instead. There are no exercises, because I have found those to be boring. The best, I think, is to work through the derivations with pad and pencil. That process also is a good way for the reader to check my work.

On the subject of checking this work, I am very grateful to Dr. Francis Bostick, who graciously did a thorough review of not only the math, but also the formatting of this booklet. I think he caught all the errors, but fear that I may have missed some in the incorporation of his notes; therefore, any errors that remain are mine.

Transformations

Table of Contents

Introduction..1
Overview..3
The Math Tools..4
    The Euler Relationship..4
    Series Representations...4
Phasors, Phase-Variables and Phase-vectors..9
    Phasors..9
    Phase-variables..10
    Phasor and Phase-variable arithmetic..13
    Phase-vectors..16
        The Exponential Definition of an Ellipse..17
    Space-vectors..19
The Transformations...23
    The General Transformation..23
    Clarke Transformation...31
    Park Transformation..38
    Fortescue...40
        The Fortescue Definition..40
        The Phasor Demonstration..41
        ... but what have we done, here?...43
        The Phase-variable Demonstration..45
Conclusion...57

Transformations

Transformations

Introduction

It is customary to model poly-phase systems with separate sources, transmission systems and loads for each phase. Although these models may include coefficients of coupling between phases that describe interaction, we will presume that is not the case in this discussion. Mathematically these systems include several equations – one set of equations for each phase. Solving these equations must be done repeatedly by digital control systems and quickly, since the time required for these solutions can limit the switching speed of the system.

By transforming the poly-phase systems into equivalent two-phase representations, we can enable faster computation and simplify the analyses for visualization and implementation of controls. If the systems are balanced, inverse transformation returns the multi-phase representation, producing signals that can be applied to controllers for each physical phase. For unbalanced systems, however, the inverse transformations from two to three or more phases are not unique. There are many sets of poly-phase voltages that can produce a given motor flux, so an inverse transformation may not (and probably will not) correctly represent the influence of an input imbalance. It is possible, however, to accurately represent unbalanced systems with a collection of sets of balanced components, then reversibly operate on the new, balanced, equations with the transformations that have been developed for balanced systems. Control systems can then operate on these equations as though they directly describe the actual drive systems.

In this booklet we will show various representations of signals and derive some of the transformation methods often used in Power Electronics. The tools we rely on most often within this paper are the complex exponential notation and the corresponding trigonometric representations. Some readers do not regularly use these forms, so we begin with a look at the exponential and trigonometric relationships. Once readers have developed an intuitive understanding of these notations, we are easily able to demonstrate the elegance and simplicity of the transformations for poly-phase systems.

It is important to note that as we derive our coordinated set of relationships, we may deviate from the definitions presented in various other writings. While studying this booklet, use the definitions presented here. We will try do show these differences as they arise.

Transformations

Transformations

Overview

After establishing a set of mathematical tools, we will begin our study with a steady state Phasor description of three voltages, currents or fluxes that make up the system. Then we learn that more information can be presented using time-varying variables that represent the values of the parameters at every instant in time. Finally, we look at a vector representation that includes the spatial orientation of, for example, flux in a motor. The following development is done using three-phase examples, but some of the work is applicable to higher order poly-phase systems as well.

We use math transformations on these vectors (we will call them "Phase-vectors" to distinguish them from Phasors)[1] to move between an array of three or more such vectors and alternative two-vector descriptions. The resultant (or vector sum) of the two-vector results of the transformations is the same as for the poly-phase sums; and the transformations are reversible for balanced systems. We name the three-vector description *abc* and the two-vector version *αβ*. The α and $j\beta$ vectors are plotted on a two-dimensional complex plane so that complex (but that doesn't mean complicated) math can be used to do modeling and analysis. This simplification is known as the Clarke transformation and we will demonstrate its application to Phase-vectors.

A second transformation is used to change the fixed frame of reference of the two-dimensional *αβ* model to a rotating reference frame with a speed matching, for example, the rotation of the resultant field in a machine. This new description is called the *dq* (direct–quadrature) model, and is produced by the Park transformation. Again, these transformations with their inverses apply to balanced systems of voltages, currents or flux fields, and can be used to describe any of these parameters.

We will explore a more nearly complete Phasor model provided by a relationship that was produced in 1918 by C. L. Fortescue, in which an *un*balanced set of three- (or n-) Phasors can be represented by three different sets of components, each of which are called symmetrical (that is, balanced). This we will follow by describing a process similar to the Fortescue method, applied to Phase-variables.

---

[1] *Some people seem to use the words Phasor and Vector interchangeably. We use Phasor to mean a fixed (on the page) line representing the magnitude of a signal and the electrical phase angle (a displacement in time) with respect to a given reference ($V\angle\theta$). Our term Phase-vector adds instantaneous information, and although it still has a fixed direction on the page, it is now a spatial displacement. The Phase-vector varies, changing both sign and magnitude in time. We will learn more about this in the sections on Phasors and Phase-vectors. Additionally, we use the term Phase-variable to mean simply the time varying value of a variable, as would be seen with an oscilloscope.*

Transformations

The Math Tools

We will derive each of the relationships above, to show how they may have been developed and how they work together. We will use complex math for the derivations because of the simplicity it affords, and we will also be using well-known trigonometric identities. To provide a bit more background, we precede our derivations with a review of the Euler relationship, in which these trigonometric functions are represented in complex exponential notation. Engineers have all seen and played with this transformation, but may not have a good intuitive understanding of the way $r = e^{j\omega t}$ traces a circle in the complex plane.

The Euler Relationship

Using the series representations of both the trigonometric and exponential functions, we will see how Leonhard Euler may have found his well known relationship.[2]

We want to show that the trig and exponential descriptions produce the same image on their respective reference frames. We start, though, with a look at the makings of the cosine, sine and exponential.

Series Representations

Joseph Fourier determined that any wave shape can be represented as the sum of various sinusoids. The exponential form of each of these waveforms is also available. In particular, Brook Taylor[3] found that these exponential terms can be arranged in a series that when summed, comprise a representative differential equation for the curve. The Taylor series is defined this way:

$$f(x) = f(a) + (x-a)f'(a) + \frac{(x-a)^2}{2!}f''(a) + \frac{(x-a)^3}{3!}f'''(a) + \cdots + \frac{(x-a)^n}{n!}f^{(n)}(a) + \cdots \quad (1)$$

(Here the prime infers the derivative with respect to *x*.) Solving this series at $a = 0$ and for *f(x)* equal to cosine, sine and exponential functions isn't difficult and produces the following sets:

---

[2] According to the Science Fair Project Encyclopedia, Euler rediscovered, in 1748, the relationship described by Roger Cotes in 1714. Apparently neither of these guys noticed the geometrical implications of the formula. Caspar Wessel did that, something like fifty years later.

[3] There is a nice discussion of these series at 'Taylor series. (2017, May 8). In Wikipedia, The Free Encyclopedia. Retrieved 20:23, May 8, 2017, from https://en.wikipedia.org/w/index.php?title=Taylor_series&oldid=779392792'; but another good reference is [4], as well as your calculus book.

Transformations

$$\cos x = 1 - \frac{x^2}{2!} + \frac{x^4}{4!} - \frac{x^6}{6!} + \cdots$$

$$\sin x = x - \frac{x^3}{3!} + \frac{x^5}{5!} - \frac{x^7}{7!} + \cdots \qquad (2)$$

$$e^x = 1 + x + \frac{x^2}{2!} + \frac{x^3}{3!} + \frac{x^4}{4!} + \frac{x^5}{5!} + \frac{x^6}{6!} + \frac{x^7}{7!} + \cdots$$

Figure 1 is an interesting chart displaying the plots of each of the terms of the series above, and also the sums of the respective terms for the cosine, sine and exponential. The bold sections of the plots of the three-term sums clearly show the way the functions are developed by the sums. This chart required nine series terms, and so we see that many terms are needed to produce the transcendental functions over a large range.

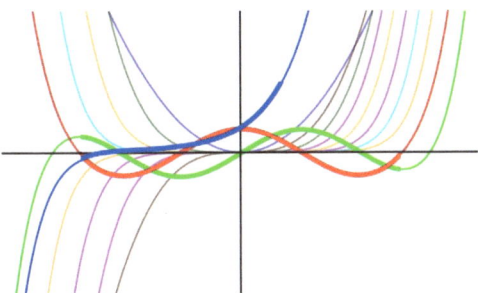

Figure 1: Taylor Series Terms and Sums

Euler evidently noticed that the sum of terms in the series expansion for the complex exponential (e^{jx}) are the same as the sum of the series for both cos x and $j$sin x, if $j = \sqrt{-1}$. So, by replacing x with jx in the representation for the exponential, we produce:

$$e^{jx} = 1 + jx - \frac{x^2}{2!} - \frac{jx^3}{3!} + \frac{x^4}{4!} + \frac{jx^5}{5!} - \frac{x^6}{6!} - \frac{jz^7}{7!} + \cdots \qquad (3)$$

Sure enough, these are exactly the sum of the cosine terms and j times the sine terms, so we can rewrite (3), separating the real and imaginary terms:

$$e^{jx} = \left[1 - \frac{x^2}{2!} + \frac{x^4}{4!} - \frac{x^6}{6!} + \cdots \right] + \left[jx - \frac{jx^3}{3!} + \frac{jx^5}{5!} - \frac{jz^7}{7!} + \cdots \right] = \cos x + j \sin x \qquad (4)$$

according to the series definitions.

Transformations

We can plot the real and imaginary parts of (4) on perpendicular axes. By doing this we are giving the two terms on the far right side of the equation properties of vectors. We let the j indicate the terms to plot against a vertical axis, perpendicular to a real-vector horizontal axis. This is the same as saying that applying the j rotates the vector by 90°. [4] Now, if we give x an angular value, say $x = \psi$, we can see that $e^{j\psi} = \cos\psi + j\sin\psi$ represents a unit vector at ∠ ψ.

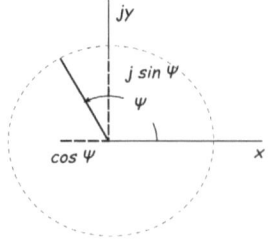

Figure 2: The complex plane

The j is, again, just a way of saying "Plot this term on the vertical axis!" It allows us to keep track of two equations at the same time, or to plot two-dimensional information with a single equation. As ψ increases over time (that is, by letting $\psi = \omega t$),

$$e^{j\omega t} = \cos\omega t + j\sin\omega t \qquad (5)$$

will trace a unit circle on the x, jy plane, as time progresses.

Here is another thing to look at while considering the Euler relationship. Just as we can define the exponential form in terms of the trig functions, we can do the reverse and define the trig functions using the exponential. First, look at what happens when we reverse the direction of rotation in (5):

$$e^{j(-\omega t)} = \cos(-\omega t) + j\sin(-\omega t) = \cos\omega t - j\sin\omega t \qquad (6)$$

Because $\cos(-\omega t) = \cos(\omega t)$, only the sign of the second term changes when we change the sign of the exponent. The equation for $e^{-j\omega t}$ is the conjugate of that for $e^{j\omega t}$, and gives us the second of two equations that we can solve for $\cos(\omega t)$ and $\sin(\omega t)$. Adding (5) to (6) we get:

$$e^{j\omega t} + e^{-j\omega t} = (\cos\omega t + j\sin\omega t) + (\cos\omega t - j\sin\omega t) = 2\cos\omega t \qquad (7)$$

or, upon re-arranging terms,

$$\cos\omega t = \frac{e^{j\omega t} + e^{-j\omega t}}{2} \qquad (8)$$

Similarly, if we subtract (6) from (5) we find that

---

[4] ... So we have created the definition: $j = e^{j\pi/2}$.

Transformations

$$\sin \omega t = \frac{e^{j\omega t} - e^{-j\omega t}}{j2}. \tag{9}$$

Now look at the effect of multiplying $e^{j\omega t}$ by some other exponential, like $e^{j\alpha}$:

$$e^{j\alpha} e^{j(\omega t)} = e^{j(\omega t + \alpha)} = \cos(\omega t + \alpha) + j\sin(\omega t + \alpha) \tag{10}$$

We have rotated the line segment by the angle α (ωt replaced with $\omega t + \alpha$), by multiplying with the unit exponential, $e^{j\alpha}$. It is also valid to write this using a combination of both exponential and trigonometric notation:

$$e^{j\alpha}(\cos(\omega t) + j\sin(\omega t)) = \cos(\omega t + \alpha) + j\sin(\omega t + \alpha) \tag{11}$$

Again, the exponential multiplier rotates the expression to which it is applied by angle α.

Here is a major opportunity for confusion! $e^{j\alpha}\cos(\omega t)$ is **not** $\cos(\omega t + \alpha)$ (even though (11) is true! Let us take another quick look at rotation of the sine and cosine functions by our exponential rotator. It seems easy, after looking at (10) and (11), to write

$$e^{j\alpha}\cos(\omega t) = \cos(\omega t + \alpha) \qquad \text{... but \textbf{THIS IS WRONG!}}$$

and can lead to trouble if we do it. To confirm this, we write the equation above in our exponential notation:

$$e^{j\alpha}\cos\omega t = e^{j\alpha}\left(\frac{e^{j\omega t} + e^{-j\omega t}}{2}\right) = \frac{e^{j(\omega t + \alpha)} + e^{-j(\omega t - \alpha)}}{2}. \tag{12}$$

This is certainly not the same as the time-shifted cosine! As ωt varies here, the two terms in the sum describe the loci of two counter-rotating line segments. Because of the signs of α within the arguments, the angle of each is advanced by α in the same (here counter-clockwise) direction. When we add these vectors, we can see that although the *time*-phase angle of the co-sinusoidal variation along the line is not changed, the direction of the line itself in *space* is rotated by the angle α.

Didn't we just do this same evil thing with each of the trig terms in (11)? Answer: yes! So why is (11) OK? Well, let's work it out and see what happens. If we rotate each of the sinusoids separately, they look something like (12):

Transformations

$$e^{j\alpha}\cos\omega t = e^{j\alpha}\left(\frac{e^{j\omega t}+e^{-j\omega t}}{2}\right) = \frac{e^{j(\omega t+\alpha)}+e^{-j(\omega t-\alpha)}}{2}$$

$$e^{j\alpha}\sin\omega t = e^{j\alpha}\left(\frac{e^{j\omega t}-e^{-j\omega t}}{j2}\right) = \frac{e^{j(\omega t+\alpha)}-e^{-j(\omega t-\alpha)}}{j2} \qquad (13)$$

The confusion is resolved when we put these back into (11) (along with the j for the sine):

$$e^{j\alpha}(\cos(\omega t)+j\sin(\omega t))$$
$$=\frac{e^{j(\omega t+\alpha)}+e^{-j(\omega t-\alpha)}}{2}+j\frac{\left(e^{j(\omega t+\alpha)}-e^{-j(\omega t-\alpha)}\right)}{j2} = e^{j(\omega t+\alpha)} \cdot \qquad (14)$$

The j factor for the sine eliminates the denominator j so that all the terms can be added. The clockwise components in this example add to zero, leaving us with a time-phase adjustment rather than a spatial rotation. It is important to keep track of the difference between the *time* and *space* rotations in these equations. They are NOT the same!

Here is one final equivalence that will come in handy. In the literature (for example [3]), we find that people use another definition: $a = e^{j\gamma}$ (γ is here $2\pi/3$); then $a^2 = e^{j2\gamma} = e^{-j\gamma}$ and the Phasor

$$\underline{V} = a^0 V_a + a^1 V_b + a^2 V_c = V_a + aV_b + a^2 V_c \qquad (15)$$

(Underbars are suppressed, here. Notice that V_a, V_b and V_c are magnitudes, not Phasors in this equation, but the set V_a, aV_b, a^2V_c are Phasors.) This notation is just another way to write our exponential rotation, here again used to advance the position of the three-phase signals above on the Phasor diagram. We mention this notation because we will use it in our section on transformations, and for the Fortescue work (as he did). Also, you will see it elsewhere in the literature. Keep in mind the simple equivalence: $a = e^{j\gamma}$ (again, γ is our $2\pi/3$).

All of this notation and its interpretation is important to the rest of this paper, so try to be comfortable with it by following the previous demonstrations with your pencil and paper. Further review is available in on-line encyclopedias as well as in your college texts.

Transformations

Phasors, Phase-Variables and Phase-vectors

Now we have the trigonometric and exponential notations as math tools that we can use to explain and work with our electrical relationships. We can also move between one representation and the other. Our goal in this booklet is to create descriptions of the fields and other quantities within a motor and power system to use as models of the system elements. To represent the fluxes (or voltages or currents) we will present three methods. We call one representation Phasors, which are shortcut versions of the second representation: Phase-variables. The third method we will call Phase-vectors. We will see that these three methods of describing parameters have similarities on paper, but are actually quite different. Let's look at the differences between these descriptions.

Phasors

First, a review of Phasors: A Phasor describing a parameter of the system is a shorthand (and therefore incomplete) representation that allows us to do loop and node calculations for ac circuits without concerning ourselves with the instantaneous values of the underlying sinusoidal functions. A Phasor looks like a vector in that it has a magnitude and a phase angle, but the angle does not describe a direction of action for its parameter; hence a Phasor is not a vector. The phase angle in a Phasor represents the *relative* electrical displacement in *time* between the similar waveforms of voltages or other parameters.[5] So, a drawing of the Phasor has magnitude and phase angle physically fixed on the page, although the underlying parameter (flux, voltage, etc.) is changing in time. The magnitude we choose to use may be the RMS or the peak value of the parameter.[6] A position can be chosen as the reference angle; perhaps we call the *a*-phase angle for the voltage of a poly-phase system zero, for example. Then, other system Phasors are defined with respect to that reference angle. The Phasor does not include frequency information, but presumes that the system is made up of variables of the same (single) frequency.

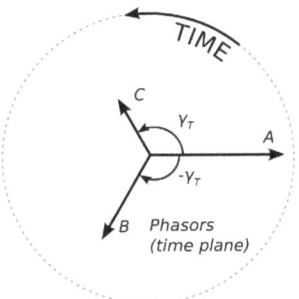

Figure 3: Phasor Set

A set of Phasors then, describes the relative phase relationships (time delays between waveforms) and magnitudes of the corresponding parameter set, on a two-dimensional chart. Here is one way we can write the Phasor description of a three-phase set of voltages with a symmetrical phase distribution. Uppercase letters, underlined, are our standard notation for Phasors. (Here we also demonstrate engineering notation on the right side of the equations):

---

[5] *This may be a good time to review page 7, re: rotations in time vs. in the spatial domain.*
[6] *Usually RMS, since calculations such as power (P = IV*) work without adjustment when using RMS.*

Transformations

$$\underline{\Phi}_a = A \angle 0°$$
$$\underline{\Phi}_b = B \angle -120°$$
$$\underline{\Phi}_c = C \angle -240° = C \angle 120° \qquad (16)$$

An offset of -240°, takes us clockwise to the same position as +120° takes us counter-clockwise.

So, the Phasors are drawn on a polar chart, in relative positions that represent angular displacements in *time* from a reference angle of our choosing. They have lengths representing either the RMS or peak values of their respective time-varying parameter – such as the A of $A\sin(\omega t)$. If the Phasors are a balanced set, their sum is always zero; but there are other sets of Phasor values that will also add to zero.

We can also write an evenly distributed Phasor representation using our exponential notation:

$$\underline{\Phi}_a = A e^{j0}$$
$$\underline{\Phi}_b = B e^{-j\gamma_T}$$
$$\underline{\Phi}_c = C e^{-j2\gamma_T} = C e^{j\gamma_T} \qquad (17)$$

in which we have used $\gamma_T = 2\pi/3$ for the offsets in time. We have again chosen the A term as the reference signal, and the exponential terms position the B and C Phasors in rotation on the page as we learned earlier.[7]

Phasors have no spatial orientation at all; however later in this note we will apply our rotator to Phasors in the time plane to do some more magic. Should we move the set of Phasors in the positive (CCW) direction, note that they would pass a reference axis in ABC order if we define the Phasors' positive ABC sequence to be clockwise. Figure 3 presents a Phasor set so that you can visualize this rotation, and see the reasoning for this definition.

Next we will add the time dependence to the mathematical Phasor description to produce Phase-variables; then compare Phasors and Phase-variables. With that done, we can come back to the Phasors with a quick look at doing arithmetic for network analyses.

Phase-variables

Phase-variables include the Phasor data plus the time dependency of the signal including details of the continuous change in value during the cycle of machine operation that were omitted with Phasors.

To demonstrate the meaning of the Phase-variable we begin by writing a time-dependent description of the variables we used Phasors to describe above. Here we are

---

[7] Here ABC may be expected to mean peak values. In the literature, however, they may be either RMS or peak. This is just a way to write the Phasor, and is the notation we will use in this document.

Transformations

using Φ_{abc} instead of ABC for the scalar magnitudes:

$$\begin{aligned}\varphi_a &= \Phi_a \cos(\omega t) \\ \varphi_b &= \Phi_b \cos(\omega t - \gamma_T) \\ \varphi_c &= \Phi_c \cos(\omega t + \gamma_T)\end{aligned} \quad (18)$$

Φ_{abc} are peak values for Phase-variables.

Phase-variables are useful in modeling or observing the voltage, current or flux in a motor, and can describe signals of different frequencies and interactions between these signals (such as modulation) within a system. These are the signals we can display with an oscilloscope, as seen in Figure 4.

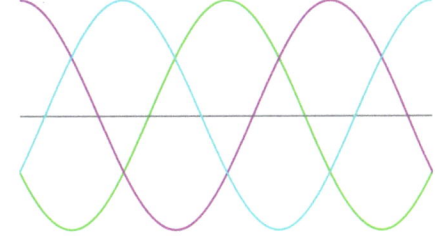

Figure 4: Phase-variables

As we have seen, we can represent these parameters as Phasors on a time plane if they share a common single frequency. Phasor notation uses only the magnitudes and the angular displacements for the offsets in time between the phases.

If this is a balanced set, these Phase-variables add to zero like the Phasors at any instant of time, even though they are changing. This variation is the way Phase-variables differ from Phasors – they do change with time, as described by (18) – busily varying in length, usually as sinusoids; but they do not describe any rotation. The arrows in Figure 5 illustrate the Phase-variables as produced by sets of windings, shown at $t=0$.

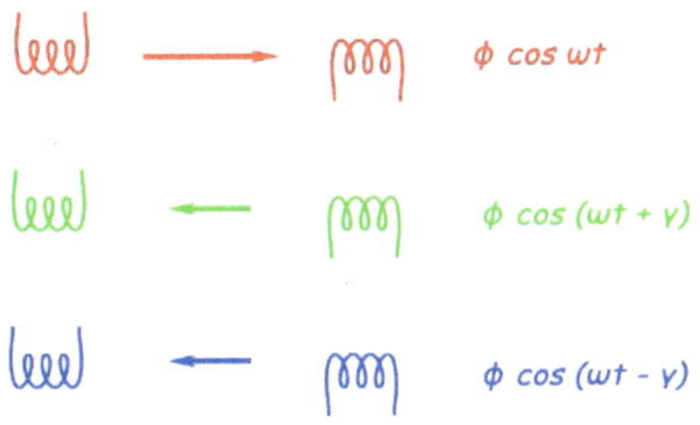

It will help us see the difference between Phasors and Phase-variables if we write one in terms of the other mathematically. We can write the Phase-variable description of (18) in exponential form:

Figure 5: Phase-variables as Flux in Windings

$$\begin{aligned}\varphi_a &= \frac{\Phi_a}{2}\left(e^{j\omega t} + e^{-j\omega t}\right) \\ \varphi_b &= \frac{\Phi_b}{2}\left(e^{j(\omega t - \gamma_T)} + e^{-j(\omega t - \gamma_T)}\right) \\ \varphi_c &= \frac{\Phi_c}{2}\left(e^{j(\omega t + \gamma_T)} + e^{-j(\omega t + \gamma_T)}\right)\end{aligned} \quad (19)$$

The frequency dependent sums on the right of (19) can now be written in terms of

Transformations

the Phasors by factoring out the time independent terms like this one for φ_b:

$$\varphi_b = \frac{\Phi_b}{2}[e^{j\omega t}\, e^{-j\gamma_T} + e^{-j\omega t}\, e^{j\gamma_T}] = \tfrac{1}{2}[\underline{\Phi}_b e^{j\omega t} + \underline{\Phi}_b^{*} e^{-j\omega t}]. \qquad (20)$$

This is the relationship between Phasors and Phase-vectors. Remember $\underline{\Phi}$ infers the Phasor, which in (20) is

$$\underline{\dot{\Phi}}_b = \Phi_b e^{-j\gamma_T}. \qquad (21)$$

This derivation has used the peak values of the Phase-variables, and if RMS values are used, the factor of ½ disappears. So we have:

$$\underline{\Phi}_b \stackrel{\text{def}}{=} \Phi_b e^{-j\gamma_T} \quad \text{(by definition)} \qquad (22)$$

This shortcut notation seems to enable adding a frequency term in a Phasor argument by multiplication by a rotator, as may be seen in the literature. This is however not in keeping with the usage in this document, since it may add confusion! – there is no frequency information in our Phasor![8]

---

[8] *Remember page 7.*

Transformations

Phasor and Phase-variable arithmetic

We have seen that to be complete, each Phase-variable description must comprise both its time dependent term and its Phasor; and also their conjugates. We have seen that authors may, however, use only the Phasor, dropping the conjugate, the time dependent exponentials and the factor of ½ presented in (20) altogether; then do their Phasor arithmetic. We can show that this is fine, but it does add another confusion feature when we are trying to see just how all this fits together! We can compare the math for the two representations to add more clarity.

First, look at Phasor multiplication: $IV^* = Ie^{j\sigma} \cdot Ve^{-j\delta} = I \cdot Ve^{j(\sigma-\delta)}$. Pretty easy – just multiply the (RMS) magnitudes and add the angles represented by the exponents. Notice that one of the multiplicands is a conjugate, and see below.

Now look at the product of two Phase-variables, using peak values for the magnitudes:

$$\begin{aligned} s &= I \cos(\omega t + \sigma) \, V \cos(\omega t + \delta) \\ &= \frac{I}{2}[e^{j\omega t}e^{j\sigma} + e^{-j\omega t}e^{-j\sigma}] \frac{V}{2}[e^{j\omega t}e^{j\delta} + e^{-j\omega t}e^{-j\delta}] \\ &= \frac{IV}{4}[e^{j(2\omega t+\sigma+\delta)} + e^{j(\sigma-\delta)} + e^{j(-\sigma+\delta)} + e^{-j(2\omega t+\sigma+\delta)}] \end{aligned} \quad (23)$$

Now we have two sets of conjugate exponentials: one pair counter-rotating at twice the frequency of the voltage and current and with phase shift equal to the sum of both; another pair at plus and minus the fixed angular difference between the I and V signals (which is a constant). We can write these back into trig form:

$$s = \frac{IV}{2}[\cos(2\omega t + \sigma + \delta) + \cos(\sigma - \delta)]. \quad (24)$$

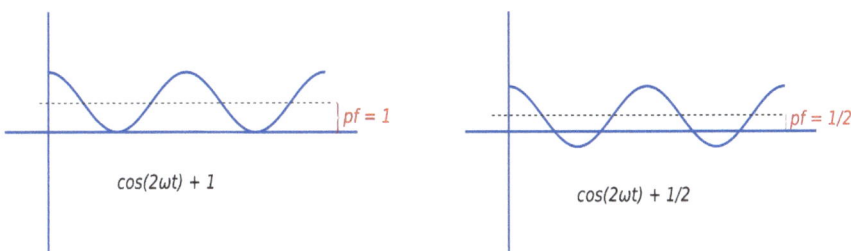

Figure 6: Unity PF   Figure 7: PF = 1/2

See that the time-dependent term and the factor of two both do show up in the *peak-value* Phase-variable product. This is the actual equation for the product of two same-frequency sinusoids for which the *(RMS)* Phasor is a *shortcut*. If we average the time-dependent term, the result is always zero, leaving ½ *I V* times the power

Transformations

factor — the same as in Phasor products (for peak values). Of course if there is no phase difference between I and V, the power factor cosine is unity, as we expect.[9] Figures 6 and 7 show the power wave-forms for unity Power Factor and PF = ½.

Starting again with (23), we can write, changing to the (*RMS*) Phasors:

$$s = \frac{I\,V}{4}[e^{j(2\omega t+\sigma+\delta)} + e^{j(\sigma-\delta)} + e^{j(-\sigma+\delta)} + e^{-j(2\omega t+\sigma+\delta)}] \quad (Peak\ values)$$

$$= \frac{1}{2}[\underline{I}\,\underline{V}\,e^{j2\omega t} + \underline{I}\,\underline{V}^* + \underline{I}^*\underline{V} + \underline{I}^*\underline{V}^*e^{-j2\omega t}] \quad (RMS\ Phasors) \tag{25}$$

Observing that the time variation average is zero, write the power Phasor:

$$\underline{S} = \frac{1}{2}[\underline{I}\,\underline{V}^* + \underline{I}^*\underline{V}] \ . \tag{26}$$

Leaving only a small exercise for the reader, we see pretty easily that

$$\underline{S} = \frac{1}{2}[IV(e^{j(\sigma-\delta)} + e^{-j(\sigma-\delta)})] \quad (again,\ RMS). \tag{27}$$

Now I and V are RMS magnitudes. Looking at the Phasor's components and their sum (in Figure 8), we find that the real parts are equal, while the conjugates include the opposite sign. The sum is then just the sum of the two real parts, which can of course be written as two times the real part of either term. That is:

$$\underline{S} = \underline{I}\,\underline{V}^* + \underline{I}^*\underline{V} \text{ or } 2\Re[\underline{I}\,\underline{V}^*] = 2\Re[\underline{I}^*\underline{V}]$$
$$= I\,V\cos(\sigma-\delta) \quad (that\ is\ IV_{RMS}*P.F.) \tag{28}$$

which is the Phasor product we see in the literature. The Phasor *product* has no direction on the time plot, which pleases our intuition since it is a different frequency than the Phasors from which it is derived, and cannot be fixed on a polar time plot with them.

For a poly-phase system, we can do another simplification for Phasors, which is called a per-phase calculation. For this we use only one phase for the

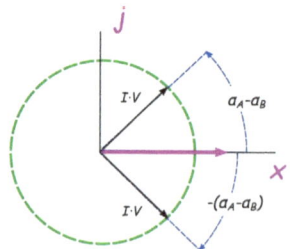

Figure 8: Phasor Elements

---
[9] You do recall that PF is defined as the cosine of the difference in phase angle between I and V?

Transformations

arithmetic and recognize that when we're done with calculations, we must correct the result. For example for a three-phase system, after a per-phase calculation of the *IV* product to determine power, we must triple the result. We formally produce this representation by using our gamma rotator to align each Phasor angle with a reference (zero angle), and then average the magnitudes. In matrix form, we can multiply and average by dividing by three for three-phases:

$$\frac{1}{3}[e^0 \; e^{j\gamma} \; e^{-j\gamma}] \cdot [S \; Se^{-j\gamma} \; Se^{j\gamma}]^T = \frac{1}{3}[S + S + S]e^{j0} = S_{avg} \qquad (29)$$

This gives us the power per phase. This Phasor procedure will, in fact, be used again later in this presentation.

Transformations

Phase-vectors

Phase-vectors add still more information to the Phase-variables' description. They add a phase displacement from a reference angle in *space*. The idea of phase displacement in space is easily visualized for vectors representing flux – three windings for a three-phase motor will be displaced physically around its stator at 120° intervals.

The sum of a balanced set of fluxes (represented as Phase-variables, by measuring the values individually with an oscilloscope as in Figure 4), will still be zero like the sum of Phasors; however, these same values will produce a rotating resultant flux in a motor if we arrange them in a symmetrical spatial distribution around the stator. (This turns them into Phase-vectors.) To accomplish this physical angular distribution mathematically, we use our exponential rotator with a *spatial* γ_s. The arrows over the exponentials infer *spatial* rotation, creating vectors from the Phase-variable representation.

$$\vec{\varphi}_a = \overrightarrow{e^{j0}} \varphi_a = \overrightarrow{e^{j0}} \Phi_a \cos(\omega t)$$
$$\vec{\varphi}_b = \overrightarrow{e^{j\gamma_s}} \varphi_b = \overrightarrow{e^{j\gamma_s}} \Phi_b \cos(\omega t - \gamma_T) \qquad (30)$$
$$\vec{\varphi}_c = \overrightarrow{e^{j2\gamma_s}} \varphi_c = \overrightarrow{e^{j2\gamma_s}} \Phi_c \cos(\omega t + \gamma_T) = \overrightarrow{e^{-j\gamma_s}} \Phi_c \cos(\omega t + \gamma_T)$$

Figure 9 is a diagram of a motor field. The three windings introduced in Figure 5 are positioned around the center by the spatial rotator coefficients in (30).

Here is an important consideration: As we noticed earlier, Phasors, although they look like vectors drawn in the time domain, are *not* technically vectors. Phasors behave as scalars in space – the Phasor's delay in *time* has no effect on the position of the Phase-vector in *space*. (Conversely, when we plot in the space domain, as in a technical drawing of motor windings, the position of the windings has no effect on the value or phase angle of the voltages.) This, too, will be an important consideration later in this booklet.

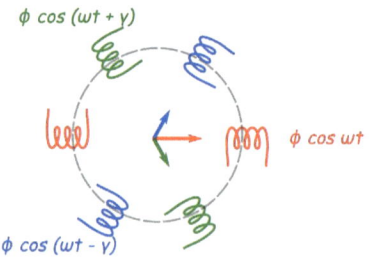

Figure 9: Physical view of Phase-vectors

After *combining* the time and space displacements, however, the vector sum of a set of distributed Phase-variables (which are therefore Phase-vectors) is a resultant vector, not zero. This will lead to our definition of a Space-vector in the next section.

Looking at Figure 10, the black lines on the left are *Phasors* that represent a parameter set (they need not be of equal length). The red and green lines on the right are two pictures of equally distributed (*120°*) Phase-*vectors* at two different times, corresponding to $\omega t = 2\pi$ and $\omega t = \pi/2$.

Transformations

Each stationary Phasor on the left of Figure 10 is shown as a magnitude and an angle with respect to a reference angle – each represents the magnitude and *electrical* phase angle of a signal while saying nothing about *physical* alignment or frequency. For a balanced set, the Phasors must add to zero. Remember that for Phase-variables, the instantaneous value of each varies, although the polar displacements in space are not assigned. These Phase-variables also add to zero in a balanced system. The Phase-vectors in the right half of Figure 10 represent two different freeze-frames of the continuously varying vector lengths. The three unit-magnitude Phase-vectors in this balanced example always add to a resultant that has a constant magnitude of 1.5, and this resultant can rotate in time. Figure 10 is important to recognizing the differences between Phasors and Phase-vectors, and is therefore worthy of careful study. Remember especially that Phasors are drawn in a *time* plane and Phase-vectors are plotted in *space*!

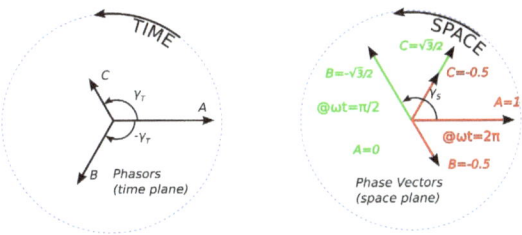

Figure 10: Compare Phasors and P-vectors

The Exponential Definition of an Ellipse

Now let us produce another tool that will help us later on. Keep in mind that at this point in the discussion, this is a general derivation and we are discussing neither Phasors nor Phase-vectors.

We have already found that a unit circle (drawn as time advances) can be represented as the locus of

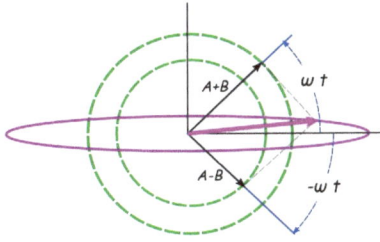

Figure 11: Parts of the Ellipse ($\alpha = 0$)

$$e^{j\omega t} = \cos \omega t + j \sin \omega t \quad \text{or} \quad \overrightarrow{e^{j\omega t}} = \overrightarrow{e^{j0}} \cos \omega t + \overrightarrow{e^{j\pi/2}} \sin \omega t \cdot \quad (31)$$

For a different sized circle,

$$Ae^{j\omega t} = A\cos \omega t + jA\sin \omega t \quad (32)$$

Pretty clearly, an ellipse can then be written as

$$A\cos \omega t + jB\sin \omega t \quad (33)$$

in which the major and minor axes of the ellipse are aligned with the real and imaginary axes, as shown in Figure 11. We can also use what we have learned to rotate that ellipse by multiplying by $e^{j\alpha}$, giving us a general representation of any ellipse centered at the origin.

Transformations

Now, again recall our exponential definitions of the sine and cosine:

$$\cos\theta = \frac{e^{j\theta} + e^{-j\theta}}{2} \tag{34}$$

and

$$\sin\theta = \frac{e^{j\theta} - e^{-j\theta}}{j2}. \tag{35}$$

Multiplying the sine by j:

$$j\sin\theta = \frac{e^{j\theta} - e^{-j\theta}}{2} \tag{36}$$

allows us to substitute these definitions into the ellipse equation, and we can rewrite our rotated ellipse in exponential terms as

$$e^{j\alpha}(A\cos\omega t + jB\sin\omega t)$$
$$= e^{j\alpha}\left[A\left(\frac{e^{j\omega t} + e^{-j\omega t}}{2}\right) + B\left(\frac{e^{j\omega t} - e^{-j\omega t}}{2}\right)\right] \tag{37}$$

This we can further reorganize into

$$e^{j\alpha} \cdot \frac{1}{2}\left[(A+B)e^{j\omega t} + (A-B)e^{-j\omega t}\right] = \frac{1}{2}\left[(A+B)e^{j(\omega t+\alpha)} + (A-B)e^{-j(\omega t-\alpha)}\right] \tag{38}$$

Look at this. It is the sum of two directed line segments describing circular loci as time advances, one with radius $A+B$ rotating in the positive direction, and one with radius $A-B$ rotating oppositely; the whole thing with phase advanced by some angle α! If $B = A$ the result is the circular trace $Ae^{j(\omega t+\alpha)}$. In Figure 11, α is zero, so the two components are aligned at $\omega t = N\pi$, on the horizontal axis. Notice that they add each time they are in line, and subtract when they oppose each other. These are the major and minor axes of the ellipse that is developed on the page as time goes by. The rotation factor, $e^{j\alpha}$, will tip the assembly, rotating the ellipse on its center.

This representation of the ellipse will also be used later on in this booklet.

Transformations

Space-vectors

The Phase-vectors are two-dimensional (since they can be modeled on a two-dimensional plane), so only two equations in two unknowns (or one complex equation) are required to fully describe each of them. We propose to write our three (or more) Phase-vector system as a set of trigonometric components with angular positions; then convert the trig forms to exponential notation and add the vectors to find the resultant. We will call this resultant a Space-vector. Although Phase-vectors usually have a spatial alignment fixed, such as by the physical windings in a machine, *the only mathematical difference between a Phase-vector and a Space-vector is that the Space-vector is the sum of all the available Phase-vectors in our machine.* A Space-vector is still two-dimensional, and we can represent it with two components, as we'll see later.

We will first describe a balanced system, so Phase-vectors for the three fluxes are:

$$\vec{\varphi}_a = \vec{e^{j0}} \Phi \cos \omega t$$
$$\vec{\varphi}_b = \vec{e^{j\gamma_s}} \Phi \cos(\omega t - \gamma_T) \qquad (39)$$
$$\vec{\varphi}_c = \vec{e^{-j\gamma_s}} \Phi \cos(\omega t + \gamma_T)$$

Here, γ_s and γ_T are the same $2\pi/3$, but the S and T indicate Spatial and Temporal shifts. Both contribute to displacement around the origin of the resultant vector sum, γ_s positioning the windings of the machine, and γ_T maintaining the electrical separation in time associated with the angular frequency. We'll use arrows over the exponentials to show that they are vectors, and have a direction of action.

Here again we recall the exponential representation of the cosine (of ωt) as

$$\cos \omega t = \frac{e^{j\omega t} + e^{-j\omega t}}{2}. \qquad (40)$$

We can use this representation to re-write the equations for the flux. As we do the multiplication in (41) we will drop the S and T suffixes, because we don't need that kind of book-keeping here – all the γ are the same magnitude and dimension and we know what each contributes. It is important, however, for us to notice that both the spatial and the temporal exponents *have the same effect on the Space-vector position in space*, since in the final sum the exponents add in the same way.

Transformations

$$\vec{\varphi}_a = e^{j0}\Phi_a\left[\frac{e^{j\omega t}+e^{-j\omega t}}{2}\right] = \Phi_a\left[\frac{e^{j(\omega t)}+e^{-j(\omega t)}}{2}\right]$$

$$\vec{\varphi}_b = e^{j\gamma_s}\Phi_b\left[\frac{e^{j(\omega t-\gamma_T)}+e^{-j(\omega t-\gamma_T)}}{2}\right] = \Phi_b\left[\frac{e^{j(\omega t)}+e^{-j(\omega t-2\gamma)}}{2}\right] = \Phi_b\left[\frac{e^{j(\omega t)}+e^{-j(\omega t+\gamma)}}{2}\right] \quad (41)$$

$$\vec{\varphi}_c = e^{-j\gamma_s}\Phi_c\left[\frac{e^{j(\omega t+\gamma_T)}+e^{-j(\omega t+\gamma_T)}}{2}\right] = \Phi_c\left[\frac{e^{j(\omega t)}+e^{-j(\omega t+2\gamma)}}{2}\right] = \Phi_c\left[\frac{e^{j(\omega t)}+e^{-j(\omega t-\gamma)}}{2}\right]$$

In the first equality of these equations, the spatial rotators are applied to the respective Phase-*variables*, which produces the three Phase-*vectors*. We can add these vectors, for example as will be done inside a motor, after spatial orientation of each Phase-vector. The time offsets add to the spatial offsets to produce the rotating field.[10] The sum of these vectors, again for this balanced case (in which $\Phi_a = \Phi_b = \Phi_c = \Phi$), is:

$$\vec{\varphi}_s = \vec{\varphi}_a + \vec{\varphi}_b + \vec{\varphi}_c = \Phi\left[\frac{3}{2}\overline{e^{j\omega t}} + \frac{\left[\overline{e^{-j\omega t}}+\overline{e^{-j(\omega t+\gamma)}}+\overline{e^{-j(\omega t-\gamma)}}\right]}{2}\right] = \frac{3}{2}\Phi\overline{e^{j\omega t}} \quad (42)$$

This works since the second (CW) terms of the expansion add to zero for the balanced case (because that dividend is three unit vectors symmetrically displaced around the origin, which must add to zero). This simple equation, $\vec{\varphi}_S = \frac{3}{2}\Phi\overline{e^{j\omega t}}$, we will call the *Space-vector* form of $\vec{\varphi}_S$ (for a balanced set of unit Phase-vectors). This form incorporates both the displacements in time and in position around the center of the chart. Changing either the time or the polar orientation arguments will affect the Space-vector angle. We have used this equation to represent flux, but of course, the notation can also be used for currents, voltages etc.

The Space-vector as developed in (41) is a very general representation, and can also be used for unbalanced systems, in which case the magnitudes Φ and general angular offsets α may even vary in time. Look next at a less specific representation of (42) with different magnitudes, and possibly different phase angles :

---

[10] *Visualize a rotary aircraft engine: The cylinders are arrayed symmetrically around the crankshaft center; but to operate, the cylinders must fire in time sequence as the timing mark advances through the spatial positions of the cylinders. To speed the engine we must increase the frequency of the firing signals. The poly-phase motor is very similar in this respect!*

Transformations

$$\vec{\varphi}_s = \vec{\varphi}_a + \vec{\varphi}_b + \vec{\varphi}_c$$

$$= \frac{[\Phi_a e^{j(\omega t + \alpha_a)} + \Phi_b e^{j(\omega t + \alpha_b)} + \Phi_c e^{j(\omega t + \alpha_c)}]}{2}$$

$$+ \frac{[\overline{\Phi_a e^{-j(\omega t + \alpha_a)}} + \overline{\Phi_b e^{-j(\omega t + \alpha_b + \gamma)}} + \overline{\Phi_c e^{-j(\omega t + \alpha_c - \gamma)}}]}{2} \qquad (43)$$

$$= \vec{\varphi}_{CCW} + \vec{\varphi}_{CW}.$$

This time the coefficients Φ_{abc} and their angles (α_{abc}) can be different, and please notice that with these differences there will generally be clockwise (CW) as well as counter-clockwise (CCW) components to the Space-vector. Here the CCW components have the same direction of rotation (or sequence) as the original set, but new vector lengths and a new resultant. The CW terms, no longer zero, now add to another magnitude and phase offset. Both the CCW and CW resultants depend on the magnitudes of Φ_{abc} and the phase terms α_{abc}. An easy way to visualize this is to draw a sketch of the six Phase-vectors with various magnitudes and small offset angles, then sketch the resultants of the two sums. Summing these counter-rotating parts produces an elliptical locus with an angular offset for the Space-vector, just as we described above in the Ellipse definition on page 17.

An important thing for us to learn now, is the effect on the counter-rotating Space-vector components if we can guarantee that the three (or N) Phase-variables always add to zero even if the system is unbalanced. (We will call this the Sum-Zero situation.) Sum-Zero can be achieved by connecting voltages in Delta, by connecting currents in Wye, or perhaps by using a processor to produce a guaranteed Sum-Zero condition. First, let's re-write the first equalities in (41) by changing our reference to phase a and placing new angular offsets in phases b and c. (In (43) we arbitrarily gave the a Phase angle a value, but we can as well use $\alpha_a = 0$.) We are making this simplification to make the next visualization easier to see. Here, if the Phase-variables (or the Phasors) for the Space-vector are Sum-Zero, those for each of the counter-rotating components must also be Sum-Zero. This is because they are conjugates, and cannot add to zero unless both are zero. Now we have

Transformations

$$\vec{\varphi}_a = \overrightarrow{e^{j0} \Phi_a} \left[\frac{e^{j\omega t} + e^{-j\omega t}}{2} \right]$$

$$\vec{\varphi}_b = \overrightarrow{e^{j\gamma_s} \Phi_b} \left[\frac{e^{j(\omega t - \gamma + \alpha'_b)} + e^{-j(\omega t - \gamma + \alpha'_b)}}{2} \right] \quad (44)$$

$$\vec{\varphi}_c = \overrightarrow{e^{-j\gamma_s} \Phi_c} \left[\frac{e^{j(\omega t + \gamma + \alpha'_c)} + e^{-j(\omega t + \gamma + \alpha'_c)}}{2} \right].$$

As before, the spatial rotators are applied to the respective Phase-*variables*, which will produce the three Phase-*vectors*. Watch when we do the multiplication: when multiplying the CCW terms, the γ add to zero in each term, and there will be a new CCW component for the Space-vector. In the CW terms, only the *sign* changes for the γ. If you sketch this operation you will see that the b and c vectors simply swap places. $\vec{\varphi}_b$ and $\vec{\varphi}_c$ still, therefore, add to $-\vec{\varphi}_a$; and the three CW vectors still add to zero, even after the spatial rotation! (We used $\alpha_a = 0$, but this works no matter the value given to α_a.) Now when we write the Space-vector, the Sum-Zero condition guarantees that there is no CW component.

$$\vec{\varphi}_s = \vec{\varphi}_a + \vec{\varphi}_b + \vec{\varphi}_c = \left[\frac{\overrightarrow{\Phi_a e^{j(\omega t)}} + \overrightarrow{\Phi_b e^{j(\omega t + \alpha'_b)}} + \overrightarrow{\Phi_c e^{j(\omega t + \alpha'_c)}}}{2} \right] \quad (45)$$
$$= \overrightarrow{\varphi_{CCW}}$$

for the Sum-Zero case. We can see that the balanced system is a special case of the Sum-Zero condition and that the locus of the Space-vector is always a circle for the Sum-Zero condition.

We can take (42) for the balanced case back to trig notation now, to continue with our visualization. We have

$$\vec{\varphi}_s = \frac{3}{2} \overrightarrow{\Phi e^{j\omega t}}$$
$$= \frac{3}{2} \overrightarrow{\Phi [\cos(\omega t) + j \sin(\omega t)]} \quad or \quad \frac{3}{2} \overrightarrow{\Phi \left[e^{j0} \cos(\omega t) + e^{j\frac{\pi}{2}} \sin(\omega t) \right]} \quad (46)$$

This is a constant magnitude vector that rotates (counter-clockwise in this example) in our complex plane as time goes by, as we expect for a balanced system. Notice that the magnitude of the Space-vector for the balanced system is always 1.5 times

Transformations

the magnitude of each of the original three Phase-vectors.

We have demonstrated methods with which to write the poly-phase system equations using consistent notation for the various representations. To avoid more possible confusion, let's acknowledge that some authors use the term Space-vector to represent what we are calling our Phase-vectors. Our Phase-vectors are aligned with respect to a spatial reference, and vary along that line. They do not have the varying angular displacement with respect to their reference that a Space-vector does have. Next we will look at some common transformations that are used in power systems engineering, and demonstrate under which conditions each is appropriate.

Transformations

The Transformations

The General Transformation

The transformations we shall discuss produce alternative representations of a poly-phase system. The new versions would produce the same Space-vector in a machine as the phase by phase system does, if the alternatives were physically implemented. These new representations can be operated upon as though the alternative equations directly model the system. For linear systems, inverse transformations of the modi-fied signals can then provide control signals for each physical phase of the system. We will consider four well-known kinds of transformation, here. First a change of representation from three or more phases to two phases; second, a rotation of the ref-erence axes. Third is a look at a combination of these that performs both operations at once. Finally, we will look at processes for producing a representation of an un-balanced system using an equivalent *set* of balanced poly-phase components. Also, we will provide for returning to the original representation in each case.

Speed and position control of poly-phase machines require separate controllers and control signals for each phase, that provide the desired effect on the Space-vector that we are ultimately controlling. We have noted that this presents us with another issue to resolve: poly-phase systems (three or more) can be said to over-define the state of a machine having a two-dimensional Space-vector[11]. For example, in a three-phase machine there are three equations describing the two-dimensional Space-vector. The solution for any two of these equations depends on the value of the third. There are an infinite number (not all, but still an infinite number) of values for the Phase-vectors that can provide the same resultant Space-vector. To try to manage the magnitude and position of the Space-vector by working with all these equations will result in chaos as the independent controllers try to bring about a particular output when they are summed. – The equations change each others' coefficients even as the solution progresses! We can handle this issue by representing the three (or more) equations as an equivalent *pair* of relationships, as we will demonstrate next.

We know from the preceding discussions that a Space-vector equation such as $\vec{\varphi}_s = \Phi_s e^{j\zeta}$ can cer-tainly be represented by two equations. These two equations describe two Phase-vectors on perpendicu-lar axes, whose vector sum is the very same as the sum of the original N Phase-vectors. (See Figure 12 for the two-vector representation.) All that is required to produce the new representation is to write the ex-ponential rotator in trigonometric form. We are using ζ for the angle between the reference axis (here x),

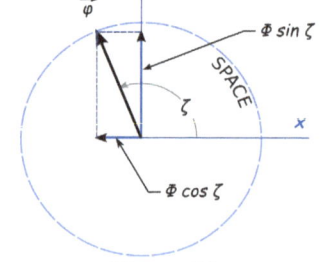

Figure 12:
*x,y representation:
Space-vector or P-vector*

---

[11] Actually, we have six equations, three each for amplitude and phase. We usually eliminate the phase equations by providing the symmetrical distribution of the windings in a ma-chine (our $\gamma = 2\pi/N$).

Transformations

and the vector in question. Although we have used the horizontal and vertical axes *x* and *y* here, this process will work for any perpendicular pair of axes, no matter their polar orientation. (Look ahead to Figures 13 and 14.)

We can represent the angular displacement of each of a set of Phase-vectors comprising a Space-vector in the same way. To extract the *x* and *y* components of $\vec{\varphi}$, for example, just put the exponential multiplier that sets its angular position into trigonometric form. So, for a single Phase-vector, $\vec{\varphi}$, with an angle ζ:

$$\vec{\varphi} = \varphi \, \overrightarrow{e^{j\zeta}}$$
$$= \varphi (\cos \zeta + j \sin \zeta)$$
$$= \vec{\varphi}_x + j\vec{\varphi}_y \qquad (47)$$

which may be written

$$\vec{\varphi}_x + \vec{\varphi}_y \text{ or } \varphi_x \overrightarrow{e^{j0}} + \varphi_y \overrightarrow{e^{j\pi/2}}.$$

Here φ, a Phase-variable, is $\Phi \cos \omega t$, ζ is the angle between the *x* reference line and the vector $\vec{\varphi}$, and $j = e^{j\pi/2}$ produces the 90° polar separation between the *x* and *y* components. (In the trigonometric form, when a component has no rotator prefix, we infer alignment with the reference axis.)

Since the Space-vector is just the sum of its component Phase-vectors, we can create a mathematical operation that will take the three- (or *N*-) phase description that produces the Space-vector to an equivalent two-phase representation. We need only extract the perpendicular components for each Phase-vector, as we did in (47),

$$\vec{\varphi}_a = \varphi_a \, \overrightarrow{e^{j\zeta_a}}$$
$$= \varphi_a (\cos \zeta_a + j \sin \zeta_a) \qquad (48)$$
$$= \vec{\varphi}_{x_a} + j\vec{\varphi}_{y_a}$$

$$\vec{\varphi}_b = \varphi_b \, \overrightarrow{e^{j\zeta_b}}$$
$$= \varphi_b (\cos \zeta_b + j \sin \zeta_b) \qquad (49)$$
$$= \vec{\varphi}_{x_b} + j\vec{\varphi}_{y_b}$$

$$\vec{\varphi}_c = \varphi_c \, \overrightarrow{e^{j\zeta_c}}$$
$$= \varphi_c (\cos \zeta_c + j \sin \zeta_c) \cdot \qquad (50)$$
$$= \vec{\varphi}_{x_c} + j\vec{\varphi}_{y_c}$$

then add the respective sets of components. The resulting Phase-variables are:

Transformations

$$\varphi_x = \varphi_a(\cos\zeta_a) + \varphi_b(\cos\zeta_b) + \varphi_c(\cos\zeta_c)$$
$$\varphi_y = \varphi_a(\sin\zeta_a) + \varphi_b(\sin\zeta_b) + \varphi_c(\sin\zeta_c)$$
(51)

Here we have kept a fixed reference frame, although changing the way we describe the phase parameters.

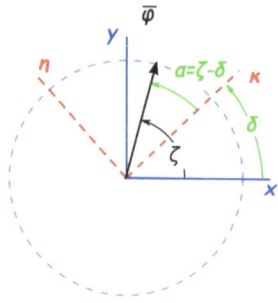

Figure 13: κ, η definition

It is often useful to base our descriptions on a rotated (or rotating) reference frame. The amplitude and polar position of the Space-vector with respect to the new frame can then be adjusted by changing two variables. (Here we must acknowledge the need for a background process operating to mathematically update the reference angle within our system equations; probably with data from a position sensor within the machine we are controlling.) We found previously that this will work for any ζ in the exponential rotator, so we can produce a rotation of the reference axes by just subtracting the angular offset δ from the original angle ζ as in Figure 13. Now we can demonstrate a general conversion with respect to any pair of (possibly rotated) reference axes. We'll again work with a three-phase system as we proceed.

First rewrite a Space-vector for a balanced system in terms of a set of three unit phase-vectors:

$$\vec{\varphi}_s = (\vec{\varphi}_a + \vec{\varphi}_b + \vec{\varphi}_c) \ . \tag{52}$$

Now we define new perpendicular components aligned with a general, (possibly even rotating as time proceeds) frame of reference with perpendicular axes we name κ and η. (See Figure 13.) We can write this two-phase representation of the Space-vector $\vec{\varphi}_S$):

$$\vec{\varphi}_s = (\vec{\varphi}_\kappa + \vec{\varphi}_\eta) \ . \tag{53}$$

Combining (52) and (53), we show the exact equivalence of the two representations

$$(\vec{\varphi}_a + \vec{\varphi}_b + \vec{\varphi}_c) \equiv (\vec{\varphi}_\kappa + \vec{\varphi}_\eta) \ . \tag{54}$$

To extract the κ and η components of $\vec{\varphi}$ in Figure 13, again put the exponential multiplier that sets the angular position of each Phase-vector into trigonometric form. Now for a single Phase-vector $\vec{\varphi}$, with an angle (defined here as $\alpha=\zeta-\delta$) to the new reference axis (κ, offset CCW by the term $e^{j\delta}$):

Transformations

$$\vec{\varphi} = \varphi \; \overrightarrow{e^{j(\zeta-\delta)}} \text{ or } \varphi \; \overrightarrow{e^{j\alpha}}$$
$$= \varphi \left(\cos\alpha + j\sin\alpha \right) \tag{55}$$
$$= \vec{\varphi_\kappa} + j\vec{\varphi_\eta}, \text{ which may be written } \vec{\varphi_\kappa} + \vec{\varphi_\eta} \text{ or } \varphi_\kappa \overrightarrow{e^{j0}} + \varphi_\eta \overrightarrow{e^{j\pi/2}}.$$

As before $j = e^{j\pi/2}$ produces the 90° separation between the κ and η components.

Next we replace ζ with ζ_{abc} (the respective polar displacements in the three Phase-vectors), and so $\alpha_{abc} = \zeta_{abc} - \delta$ for each of the three ζ, to move to the new reference axes:

$$\vec{\varphi_a} = \varphi \; \overrightarrow{e^{j(\zeta_a - \delta)}} \text{ or } \varphi \; \overrightarrow{e^{j\alpha_a}}$$
$$= \varphi_a \left(\cos\alpha_a + j\sin\alpha_a \right) \tag{56}$$
$$= \vec{\varphi_{a_\kappa}} + j\vec{\varphi_{a_\eta}}$$

$$\vec{\varphi_b} = \varphi_b \; \overrightarrow{e^{j(\zeta_b - \delta)}} \text{ or } \varphi \; \overrightarrow{e^{j\alpha_b}}$$
$$= \varphi_b \left(\cos\alpha_b + j\sin\alpha_b \right) \tag{57}$$
$$= \vec{\varphi_{b_\kappa}} + j\vec{\varphi_{b_\eta}}$$

$$\vec{\varphi_c} = \varphi_c \; \overrightarrow{e^{j(\zeta_c - \delta)}} \text{ or } \varphi \; \overrightarrow{e^{j\alpha_c}}$$
$$= \varphi_c \left(\cos\alpha_c + j\sin\alpha_c \right) \cdot \tag{58}$$
$$= \vec{\varphi_{c_\kappa}} + j\vec{\varphi_{c_\eta}}$$

When we add the components of the conjugate sets of Phase-variables, we will again have a two-dimensional representation of the same Space-vector. The magnitudes of these two components are the Phase-variables:

$$\varphi_\kappa = \varphi_a(\cos\alpha_a) + \varphi_b(\cos\alpha_b) + \varphi_c(\cos\alpha_c)$$
$$\varphi_\eta = \varphi_a(\sin\alpha_a) + \varphi_b(\sin\alpha_b) + \varphi_c(\sin\alpha_c) \tag{59}$$

With $\alpha_{abc} = \zeta_{abc} - \delta$.

So φ_κ is the length of the projection of $\vec{\varphi_s}$ onto the κ axis, and φ_η is the length of the projection of $\vec{\varphi_s}$ onto the η axis.

Above we represent each Phase-vector first as a Phase-variable with an exponential spatial rotator, then (in **bold**) as the Phase-variable with the same polar displacement

Transformations

expressed in trigonometric form. The third representation is as pairs of Phase-vectors, with rotators e^{j0} (inferred) and $e^{j\pi/2}=j$. These rotators position the Phase-variables with respect to the new reference axes. The transformation we are deriving is designed to separate the perpendicular components of each vector in this manner, and then sum the respective components of the three sets to produce the new, two-phase representation.

The preceding operations were done on a set of Phase-vectors, and now we propose an important consideration: because it is the Phase-*variable* that we can easily observe and modify with our instrumentation, it makes sense to operate on that representation for the control signals. We have gone to a lot of trouble to keep separate the temporal and spatial rotators and the signal representation upon which they operate up to now in this document. However, we found in (41) on page 20 that the polar position of the Space-vector is affected in the same way for a change in either the spatial angle or the electrical phase angle in the Phase-vector. This suggests that we can apply a rotator to the Phase-*variable* instead of the Phase-*vector* to produce a change in the polar position of the Space-vector. Now we need to justify this assertion!

We saw at (55) that the orthogonal axis extraction process works for any ζ in the exponential rotator, so we can operate directly on Phase-variables if we write them as zero-angle Phase-vectors.

So if we *assign* a direction along a reference axis to a Phase-variable, it becomes a Phase-vector with a spatial angle zero. For the Phase-variable $\varphi = \Phi\cos(\omega t + \beta)$ we can write the equivalent Phase-vector

$$\vec{\varphi} = \Phi\cos(\omega t + \beta)\overline{[\cos(0)+j\sin(0)]} = \Phi\cos(\omega t+\beta)\overline{[1+j0]} = \varphi\overline{(1+j0)} \qquad (60)$$

because $[\cos(0) +j \sin(0)] \equiv 1$ and again no prefix for the cosine infers no rotation, or e^{j0}. At (11) on page 7 we found that

$$e^{j\alpha}[\cos(\omega t)+ j\sin(\omega t)] = \cos(\omega t +\alpha)+ j\sin(\omega t+\alpha) \qquad (61)$$

for any α or any ωt. Then

$$e^{j\delta}\vec{\varphi} = \vec{e^{j\delta}}\,\varphi\overline{[\cos(0)+j\sin(0)]} = \varphi\,\vec{e^{j\delta}}. \qquad (62)$$

Mathematically we have augmented the spatial angle, but this will have the same effect on the Space-vector (according to (41)) as a change in the electrical angle. The physical system will ultimately provide spatial alignment by applying the signals to the respective windings in a machine.

We therefore continue by putting only the Phase-variables from (56)-(58) into matrix form as we give a name (T_g) to a transformation that will take us from φ_{abc} to $\varphi_{\kappa\eta}$.

Transformations

$$\begin{bmatrix} \varphi_\kappa \\ \varphi_\eta \end{bmatrix} = T_g \cdot \begin{bmatrix} \varphi_a \\ \varphi_b \\ \varphi_c \end{bmatrix}. \tag{63}$$

Filling in the blanks using (59) makes the matrix equation look this way:

$$\begin{bmatrix} \varphi_\kappa \\ \varphi_\eta \end{bmatrix} = \begin{bmatrix} \cos(\alpha_a) & \cos(\alpha_b) & \cos(\alpha_c) \\ \sin(\alpha_a) & \sin(\alpha_b) & \sin(\alpha_c) \end{bmatrix} \cdot \begin{bmatrix} \varphi_a \\ \varphi_b \\ \varphi_c \end{bmatrix}, \tag{64}$$

in which the α_{abc} in the transformation matrix are the respective offset angles between the new reference and each of the three Phase-vectors in (56)-(58). So we have defined our general transformation matrix:

$$T_g \stackrel{\text{def}}{=} \begin{bmatrix} \cos(\alpha_a) & \cos(\alpha_b) & \cos(\alpha_c) \\ \sin(\alpha_a) & \sin(\alpha_b) & \sin(\alpha_c) \end{bmatrix}. \tag{65}$$

Each of the two results in (64) comprises three Phase-variable *components* of each original phase.

To visualize this rotation of the reference frame in our mathematical model, we can simply use our exponential multiplier again; that is by adding or subtracting from the exponent of the Space-vector description we have derived above (Remember Page 19):

$$\vec{\varphi}_s{'} = \vec{\varphi}_s e^{-j\delta} = \overline{\Phi(e^{j\omega t})} e^{-j\delta} = \overline{\Phi(e^{j(\omega t - \delta)})} \tag{66}$$

$\vec{\varphi}_s{'}$ is now a description of the same vector, but with respect to the new κ, η reference frame. The negative sign on δ corresponds to a presumed positive position of the new reference axes. (See Figure 14). Now all we need to do to extract the perpendicular components is rewrite $\vec{\varphi}_s{'}$ in trigonometric form:

$$\vec{\varphi}_s{'} = \overline{\Phi(\cos\alpha + j\sin\alpha)} = \vec{\varphi}_\kappa + \vec{j}\vec{\varphi}_\eta, \tag{67}$$

in which now $\alpha = \omega t - \delta$, and j is the same as $e^{j\pi/2}$. No prefix for the cosine infers no rotation, or e^{j0}.

Transformations

Here's another approach: just as we applied the change of reference to the Space-vector, we can apply it to the entire *set* of Phase-vectors that produce the Space-vector to effect the same rotation of the system. We want to show that we can also apply it to any set of two or more Phase-*variables* before or after applying the extraction transformation. Remember the rotation works for any angle ζ, including zero. Also as we did for the multiplication in (41) we could drop the S and T suffixes in the rotators, because their effects were additive.

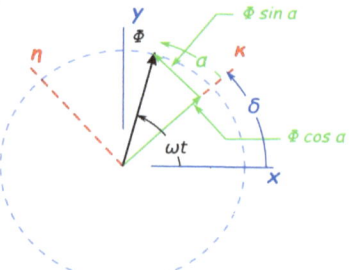

Figure 14: Extraction of κ and η Components

When we put the components into matrix form, we don't use the *j* because the two equations are treated independently by the matrix notation without it. When we use the rotator on the matrix, the angle added in the time domain has the same effect on the Space-vector as a spatial rotation.

$$\begin{bmatrix} \varphi_\kappa \\ \varphi_\eta \end{bmatrix} = e^{j\delta} \begin{bmatrix} \varphi_x \\ \varphi_y \end{bmatrix} = e^{j\delta} \left[T_g \cdot \begin{bmatrix} \varphi_a \\ \varphi_b \\ \varphi_c \end{bmatrix} \right]. \qquad (68)$$

Now the two new Phase-variables represent perpendicular Phase-vectors that have a new angular reference that we can choose by setting δ. We can adopt the new reference plane by drawing the new κ axis as our horizontal base line. (Remember that δ can be any angle, even a time dependent rotating reference; and that the angle δ in our equation must be continuously updated since it is probably changing.) Control and compensation circuitry can operate on these virtual signals as the independent paths of a two channel control network. We can control the amplitude and angular displacement of the Space-vector with respect to the new baseline by managing just the magnitudes of φ_κ and φ_η. Let's go one step further and write a general form that allows us to change the angles collectively (with δ) or individually (with α_{abc}). We write the new transformation this way, with δ replaced with $\delta + \alpha_{abc}$:

$$T_G = \begin{bmatrix} \cos(\delta+\alpha_a) & \cos(\delta+\alpha_b) & \cos(\delta+\alpha_c) \\ \sin(\delta+\alpha_a) & \sin(\delta+\alpha_b) & \sin(\delta+\alpha_c) \end{bmatrix}. \qquad (69)$$

We will put this form to good use just ahead.

> This transformation works on Phase-vectors or Phase-variables (as Phase-vectors with spatial ∡ = 0). It does *not* work on Phasors. We can demonstrate this with an example: a balanced set of Phasors represents a system with a Space-vector of fixed length, rotating about its origin with an angular position of, say, ωt. If we apply this transformation to these Phasors, the two-Phasor result must

Transformations

add to the same zero as the original three. This cannot be, unless both of the perpendicular Phasors are zero, in which case they do not well represent the non-zero Space-vector.

The inverse transformation is also important, but if we are paying attention, we note that there is no inverse for this non-square transformation matrix. (...or is there?) This we will explore further on. Let us proceed now to develop some special transformations. Again, we will apply them to Phase-variables because these are the forms with which our instrumentation will supply us.

Transformations

Clarke Transformation

Here we consider application of our general transformation to Phase-variables in some detail. Edith Clarke, known for applying mathematics to electrical power systems analyses, produced a method of transformation that takes poly-phase systems to two-phase equivalents.[1] [2] This section will explore this transformation, deriving forms that are seen in the literature today. We will start by producing the Clarke transformation for our evenly distributed three-phase system using our general transformation. This we will follow with the derivation of an inverse transformation, and a look at the way Clarke works with the three-phase system. The complex notation indicates that the real and imaginary co-ordinates of the Space-vector that results from adding the individual Phase-variable contributions can be found using sine and cosine functions to extract the components from the polar representation, as illustrated in Figure 15. Our derivation shows that the complex and *x, y* representations have identical results. The complex representation just allows us to use a single equation.

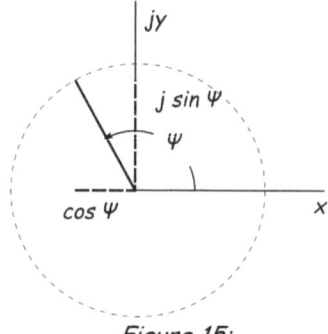

Figure 15:
The Clarke Components of $e^{j\psi}$.

We can use our general transformation by changing κ and η back to *x* and *y*, (that is setting $\delta = 0$) and α_{abc} to 0, γ and -γ, respectively. Now our general transformation becomes

$$T_C = \begin{bmatrix} \cos(0) & \cos(\gamma) & \cos(-\gamma) \\ \sin(0) & \sin(\gamma) & \sin(-\gamma) \end{bmatrix}$$
$$= \begin{bmatrix} 1 & \cos(\gamma) & \cos(\gamma) \\ 0 & \sin(\gamma) & -\sin(\gamma) \end{bmatrix} , \tag{70}$$

We can write out the two equations for a set of Phase-variables by separating the real and imaginary parts into *x* and *y* components, respectively:

$$\varphi_x = (\varphi_{ax} + \varphi_{bx} + \varphi_{cx}) = (\varphi_a \cos(0) + \varphi_b \cos(\gamma) + \varphi_c \cos(-\gamma))$$
$$\varphi_y = (\varphi_{ay} + \varphi_{by} + \varphi_{cy}) = (\varphi_a \sin(0) + \varphi_b \sin(\gamma) + \varphi_c \sin(-\gamma)) \tag{71}$$

or

$$\varphi_x = (\varphi_a + \varphi_b \cos(\gamma) + \varphi_c \cos(\gamma))$$
$$\varphi_y = (0 + \varphi_b \sin(\gamma) - \varphi_c \sin(\gamma)) \tag{72}$$

Transformations

for which, now, $\vec{\varphi}_s = \overrightarrow{e^{j0}\varphi_x} + \overrightarrow{e^{j\frac{\pi}{2}}\varphi_y} = \vec{\varphi}_x + \vec{\varphi}_y$.

This is the same flux described in (41) and (42) above, but represented with two equations rather than three.

To write the equivalence of (42) and (72) another way, we can define

$$\alpha = \varphi_x, \text{ and}$$
$$\beta = \varphi_y \tag{73}$$

then we can write

$$\vec{\varphi}_s = \overrightarrow{e^{j0}\alpha} + \overrightarrow{e^{j\frac{\pi}{2}}\beta} = \vec{\alpha} + \vec{j\beta}. \tag{74}$$

With $\vec{\varphi}_s$, again, the same resultant vector as in (42). This is the $\alpha\beta$ representation we see in other writings using these transformations.

We can now write the terms in (74) as a matrix equation,

$$\begin{bmatrix}\alpha\\\beta\end{bmatrix} = \begin{bmatrix}1 & \cos(\gamma) & \cos(\gamma)\\0 & \sin(\gamma) & -\sin(\gamma)\end{bmatrix} \cdot \begin{bmatrix}\varphi_a\\\varphi_b\\\varphi_c\end{bmatrix} = T_{C_1} \cdot \begin{bmatrix}\varphi_a\\\varphi_b\\\varphi_c\end{bmatrix} \tag{75}$$

T_{C_1} above is one form of the Clarke Transformation used in the literature.

Now, having examined the forward Clarke transformation, what about its inverse? We see in (75) that the dimensions of the transformation matrix are 2 × 3. As we discussed on page 24, this is not an invertible matrix! Why not? Well we say, in the vernacular, that the three-equation description of the three-phase (but two dimensional) system exhibits dependence. That just means that three equations are too many for the two-dimensional model, because for solutions of more than two equations for this system, the results for any one of the parameters depends on

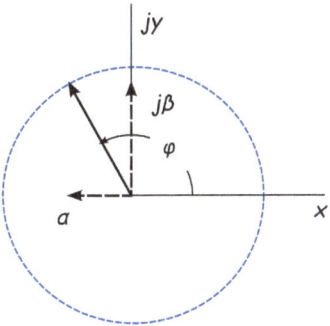

Figure 16: a,β form of Clarke

the values of the others – the solutions of the three together are not unique. Or, for three or more flux sources in a two-dimensional system, there are many sets of components that can produce the same resultant vector.

However, if we add another constraint to the definition without adding a new variable, we can have three equations for three unknowns, and therefore an invertible transformation. The constraint we will start with is the zero-sum of the Phase-vari-

Transformations

ables:[12]

$$\varphi_a + \varphi_b + \varphi_c = 0 \qquad (76)$$

This condition can be guaranteed, for example, by the delta connection of voltages or the wye connection of currents in a system.[13]

Now if we write the system for all three equations (75) becomes:

$$\begin{bmatrix} \alpha \\ \beta \\ 0 \end{bmatrix} = \begin{bmatrix} 1 & \cos(\gamma) & \cos(\gamma) \\ 0 & \sin(\gamma) & -\sin(\gamma) \\ 1 & 1 & 1 \end{bmatrix} \cdot \begin{bmatrix} \varphi_a \\ \varphi_b \\ \varphi_c \end{bmatrix} = T_{C_2} \cdot \begin{bmatrix} \varphi_a \\ \varphi_b \\ \varphi_c \end{bmatrix} = \begin{bmatrix} \varphi_x \\ \varphi_y \\ 0 \end{bmatrix} \qquad (77)$$

This transformation will still provide the same values for the *x* and *y* components, but now has an inverse, which we present here without derivation. It is seen in the literature as:

$$T_C^{-1} = \frac{1}{(1-\cos\gamma)} \begin{bmatrix} 1 & 0 & -\cos\gamma \\ -\frac{1}{2} & \frac{1-\cos\gamma}{2\sin\gamma} & \frac{1}{2} \\ -\frac{1}{2} & \frac{-1-\cos\gamma}{2\sin\gamma} & \frac{1}{2} \end{bmatrix}$$

or (78)

$$= \frac{1}{(1-\cos\gamma)} \begin{bmatrix} 1 & 0 & -\cos\gamma \\ -\frac{1}{2} & \sin\gamma & \frac{1}{2} \\ -\frac{1}{2} & -\sin\gamma & \frac{1}{2} \end{bmatrix}$$

A third equation is required for the derivation of this inverse.

Some confusion can arise here, too. We have shown that three equations are required to derive the reversible transformation, but in the literature, we will often see a two-equation Clarke transformation and two-equation inverse! This apparent lack of rigor is allowed if we require that the system be Sum-Zero. The transformation can be, and often is written without the third equation, as we did at the beginning of this section. We *infer* the Sum-Zero condition and simply disregard the third row in the

---

[12] Remember, the sum of the signed magnitudes or instantaneous Phase-variable values can be zero even if the Phase-vector sum is not!

[13] Another footnote! Actually, we could use a non-zero sum for this. Here we define $\varphi_a + \varphi_b + \varphi_c = \gamma$ (Now this is not our $\gamma = 2\pi/3$, but is an often used representation out in the real world.) Then use this equation the same way, with $\gamma/3$ assigned to each phase. The result is an **αβγ** transformation that you will also see out there. The *x* and *y* values are the same after this change, for uncoupled systems.

Transformations

transformation matrix and the third column in the inverse (once we have developed the inverse). Now, our transformations look like this:

$$T_C = \begin{bmatrix} 1 & \cos(\gamma) & \cos(\gamma) \\ 0 & \sin(\gamma) & -\sin(\gamma) \end{bmatrix} \text{ and} \quad (79)$$

$$T_C^{-1} = \frac{1}{(1-\cos\gamma)} \begin{bmatrix} 1 & 0 \\ -\frac{1}{2} & \sin\gamma \\ -\frac{1}{2} & -\sin\gamma \end{bmatrix} \quad (80)$$

for the Sum-Zero case.

Now, just to to demonstrate some of the forms of the Clarke Transformation that will be used in the Literature, we can use some trigonometric identities on the terms in the transformation for the balanced, or Sum-Zero equal distribution case. In particular, the half-angle relationship $\sin\left(\frac{\alpha}{2}\right) = \pm\sqrt{\frac{1-\cos\alpha}{2}}$ can be rewritten $2\sin^2(\alpha/2) = 1-\cos(\alpha)$, or equivalently $2\sin^2\gamma = 1-\cos(2\gamma)$.
If $\gamma = 120°$, $\cos 2\gamma = \cos \gamma$, so in that case, $2\sin^2\gamma = 1-\cos(\gamma)$. This, along with noting that $\cos \gamma = -\frac{1}{2}$, allows us to write:

$$T_C^{-1} = \frac{1}{(1-\cos\gamma)} \cdot \begin{bmatrix} 1 & 0 \\ \cos\gamma & \sin\gamma \\ \cos\gamma & -\sin\gamma \end{bmatrix} \quad (81)$$

which is the same as (80) but is another form seen in the literature.

Continuing with the same case (Sum-Zero and our γ fixed at $2\pi/3$), we can replace the trig notation with the actual values and get

Transformations

$$T_C = \begin{bmatrix} 1 & -\dfrac{1}{2} & -\dfrac{1}{2} \\ 0 & \dfrac{\sqrt{3}}{2} & -\dfrac{\sqrt{3}}{2} \end{bmatrix} \quad \text{and} \quad T_{C_2}^{-1} = \dfrac{2}{3}\begin{bmatrix} 1 & 0 \\ -\dfrac{1}{2} & \dfrac{\sqrt{3}}{2} \\ -\dfrac{1}{2} & -\dfrac{\sqrt{3}}{2} \end{bmatrix}. \qquad (82)$$

Interestingly, the (abbreviated) inverse Clarke is ⅔ of the (abbreviated) Clarke, transposed.

With very small modification of the transformation definition, we can write

$$T_C = \dfrac{2}{3}\begin{bmatrix} 1 & -\dfrac{1}{2} & -\dfrac{1}{2} \\ 0 & \dfrac{\sqrt{3}}{2} & -\dfrac{\sqrt{3}}{2} \end{bmatrix} \quad \text{and} \quad T_{C_2}^{-1} = \begin{bmatrix} 1 & 0 \\ -\dfrac{1}{2} & \dfrac{\sqrt{3}}{2} \\ -\dfrac{1}{2} & -\dfrac{\sqrt{3}}{2} \end{bmatrix}, \qquad (83)$$

or

$$T_C = \sqrt{\dfrac{2}{3}}\begin{bmatrix} 1 & -\dfrac{1}{2} & -\dfrac{1}{2} \\ 0 & \dfrac{\sqrt{3}}{2} & -\dfrac{\sqrt{3}}{2} \end{bmatrix}$$

and

$$T_{C_2}^{-1} = \sqrt{\dfrac{2}{3}}\begin{bmatrix} 1 & 0 \\ -\dfrac{1}{2} & \dfrac{\sqrt{3}}{2} \\ -\dfrac{1}{2} & -\dfrac{\sqrt{3}}{2} \end{bmatrix}, \qquad (84)$$

which are other forms you may encounter[14]. This is OK, because the transformations are by *definition*. – different versions for different uses! The ⅔ factor, for example, is often prefixed in the definitions we justified above, as we will do in the equations below. There is this, however: although one may associate the prefix with either the

---

[14] In the literature, a factor of 2/3 or √2/3 may or may not be used. Since our overlay of the complex and x-y reference frames is by definition (maybe not quite arbitrary), one can also define any scale factor desired. The result will be different only by the scale factor – angles and relative magnitudes will be preserved. We will discuss the way the 2/3 may have arisen below.

Transformations

transformation or its inverse, the value of the Space-vector comprised of the two-phase representation is the same as the original Space-vector only if the ⅔ factor is associated with the inverse!

Let's work through an example. We can write, adding a term α to represent a collective phase shift (as will be seen for reactive lead or lag in the system),

$$T_C \cdot \varphi_{abc} = \begin{bmatrix} 1 & -\frac{1}{2} & -\frac{1}{2} \\ 0 & \frac{\sqrt{3}}{2} & -\frac{\sqrt{3}}{2} \end{bmatrix} \cdot \begin{bmatrix} \Phi\cos(\omega t+\alpha) \\ \Phi\cos(\omega t-\gamma+\alpha) \\ \Phi\cos(\omega t+\gamma+\alpha) \end{bmatrix}, \text{ so} \qquad (85)$$

$$\Phi \begin{bmatrix} \cos(\omega t+\alpha)-\frac{1}{2}\cos(\omega t-\gamma+\alpha)-\frac{1}{2}\cos(\omega t+\gamma+\alpha) \\ 0+\frac{\sqrt{3}}{2}\cos(\omega t-\gamma+\alpha)-\frac{\sqrt{3}}{2}\cos(\omega t+\gamma+\alpha) \end{bmatrix} = \begin{bmatrix} \varphi_\alpha \\ \varphi_\beta \end{bmatrix} \qquad (86)$$

Then in reverse;

$$T_C^{-1} \cdot \begin{bmatrix} \varphi_\alpha \\ \varphi_\beta \end{bmatrix}$$

$$= \frac{2}{3}\begin{bmatrix} 1 & 0 \\ -\frac{1}{2} & \frac{\sqrt{3}}{2} \\ -\frac{1}{2} & -\frac{\sqrt{3}}{2} \end{bmatrix} \cdot \Phi \begin{bmatrix} \cos(\omega t+\alpha)-\frac{1}{2}\cos(\omega t-\gamma+\alpha)-\frac{1}{2}\cos(\omega t+\gamma+\alpha) \\ 0+\frac{\sqrt{3}}{2}\cos(\omega t-\gamma+\alpha)-\frac{\sqrt{3}}{2}\cos(\omega t+\gamma+\alpha) \end{bmatrix} \qquad (87)$$

$$= \frac{2}{3}\Phi \times$$

$$\begin{bmatrix} \cos(\omega t+\alpha)-\frac{1}{2}\cos(\omega t-\gamma+\alpha)-\frac{1}{2}\cos(\omega t+\gamma+\alpha) \\ -\frac{1}{2}\left(\cos(\omega t+\alpha)-\frac{1}{2}\cos(\omega t-\gamma+\alpha)-\frac{1}{2}\cos(\omega t+\gamma+\alpha)\right)+\frac{\sqrt{3}}{2}\left(\frac{\sqrt{3}}{2}\cos(\omega t-\gamma+\alpha)-\frac{\sqrt{3}}{2}\cos(\omega t+\gamma+\alpha)\right) \\ -\frac{1}{2}\left(\cos(\omega t+\alpha)-\frac{1}{2}\cos(\omega t-\gamma+\alpha)-\frac{1}{2}\cos(\omega t+\gamma+\alpha)\right)-\frac{\sqrt{3}}{2}\left(\frac{\sqrt{3}}{2}\cos(\omega t-\gamma+\alpha)-\frac{\sqrt{3}}{2}\cos(\omega t+\gamma+\alpha)\right) \end{bmatrix}$$

Transformations

$$= \frac{2}{3}\Phi \begin{bmatrix} \frac{3}{2}\cos(\omega t + \alpha) \\ \frac{3}{2}\cos(\omega t - \gamma + \alpha) \\ \frac{3}{2}\cos(\omega t + \gamma + \alpha) \end{bmatrix} = \begin{bmatrix} \Phi\cos(\omega t + \alpha) \\ \Phi\cos(\omega t - \gamma + \alpha) \\ \Phi\cos(\omega t + \gamma + \alpha) \end{bmatrix} = \begin{bmatrix} \varphi_a \\ \varphi_b \\ \varphi_c \end{bmatrix}. \tag{88}$$

So we see that this does work, and we have justified the shorthand descriptions of the transformation. (It would have worked just as well with the other prefix factors.)

We can abbreviate the description even more for the balanced condition, to demonstrate the derivation of other forms seen in print. Since γ is 120°, we can again replace the functions of γ with explicit values in (72):

$$\varphi_x = \varphi_a + \varphi_b \cos(\gamma) + \varphi_c \cos(\gamma) = \varphi_a - \frac{1}{2}\varphi_b - \frac{1}{2}\varphi_c$$

$$\varphi_y = 0 + \varphi_b \sin(\gamma) - \varphi_c \sin(\gamma) = 0 + \frac{\sqrt{3}}{2}\varphi_b - \frac{\sqrt{3}}{2}\varphi_c \tag{89}$$

then use the constraint: $\varphi_a + \varphi_b + \varphi_c = 0$ (which provides $\varphi_c = -\varphi_a - \varphi_b$, etc.) to eliminate φ_b and φ_c in the φ_x equation, and either φ_b or φ_c in the φ_y equation (so the values of only two phases must be known):

$$\varphi_x = \frac{3}{2}\varphi_a$$

$$\varphi_y = \frac{\sqrt{3}}{2}\varphi_a + \sqrt{3}\varphi_b \tag{90}$$

then

$$\begin{bmatrix} \varphi_x \\ \varphi_y \end{bmatrix} = \frac{3}{2}\begin{bmatrix} 1 & 0 & 0 \\ \frac{1}{\sqrt{3}} & \frac{2}{\sqrt{3}} & 0 \end{bmatrix}\begin{bmatrix} \varphi_a \\ \varphi_b \\ \varphi_c \end{bmatrix} = \frac{3}{2}\begin{bmatrix} 1 & 0 \\ \frac{1}{\sqrt{3}} & \frac{2}{\sqrt{3}} \end{bmatrix}\cdot\begin{bmatrix} \varphi_a \\ \varphi_b \end{bmatrix} \quad \left(=\begin{bmatrix} \varphi_\alpha \\ \varphi_\beta \end{bmatrix}\right). \tag{91}$$

This is a *very* simple representation of the Clarke transformation for balanced three-phase systems. Simple enough, in fact, that it can be implemented with a couple of op-amps!

It should now be clear that this simplified form of the Clarke transformation and its inverse is valid *if and only if the system is balanced*! This set of equations can be multiplied by a constant to change the amplitude, or ω can be changed to vary the frequency of the system. Also, we can independently control the *x* and *y* components

Transformations

of the resultant Space-vector, which could be useful in some applications.

Transformations

Park Transformation

Lets continue with a look at another of the representations we will see in the literature. Start with the shifted reference we set up in (66):

$$\vec{\varphi}_s' = \Phi\overline{\left(e^{j(\omega t - \delta)}\right)}, \qquad (92)$$

setting our previous $\zeta = \omega t$. This time we rename the movable κ and η co-ordinate axes to d (direct) and q (quadrature). If ω_S is a velocity we can define $\delta = \omega_s t - \alpha$ — remember that δ is the offset we added to shift the reference — we can rewrite the Space-vector using the notation of Figure 17 this way:

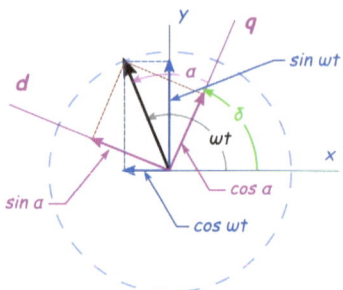

Figure 17: The Park or dq Representation

$$\vec{\varphi}_s' = \Phi\overline{\left(e^{j(\omega t - \delta)}\right)} = \Phi\overline{\left(e^{j(\omega t - \omega_s t + \alpha)}\right)}. \qquad (93)$$

Let the stator field of a motor, for example, rotate at the rate ω_S, the angular velocity of the field's Space-vector. If the reference frame of this system is made to revolve at the same rate as the moving Space-vector, the direction of that vector within the new frame will no longer change in time. This we can accomplish by setting $\omega = \omega_S$, so that

$$\vec{\varphi}_s' = \Phi\overline{e^{j\alpha}}. \qquad (94)$$

Here we have used ω_S to represent the angular velocity of the new rotation component, and α as any additional polar offset we choose. When we set $\omega = \omega_S$, and then stand upon the new rotating reference frame, the vector $\vec{\varphi}_s'$ will appear stationary to us. That means that if we know the value of ω, we can apply this reverse rotation, then work on the stationary result to set the magnitude of the Space-vector to a desired level, at an angle with respect to the new reference that we can specify. We must continually either measure or estimate the physical position $\omega_s t$ within our machine to use it in this transformation, and establish a background process to do the ongoing replacement.

However, as we noted above, to manage three Phase-variables we must represent the Space-vector using one equation for each phase before applying the transformation that produces the equivalent two phase system. The derivation of this transformation is very simple if we begin with the standard transformation we derived in (69). By letting α_{abc} equal 0, $-\gamma$ and γ respectively, and changing the sign of δ, we have, for the arrangement in Figure 17:

Transformations

$$\begin{bmatrix} \varphi_d \\ \varphi_q \end{bmatrix} = \begin{bmatrix} \cos(-\delta) & \cos(\gamma-\delta) & \cos(-\gamma-\delta) \\ \sin(-\delta) & \sin(\gamma-\delta) & \sin(-\gamma-\delta) \end{bmatrix} \cdot \begin{bmatrix} \varphi_a \\ \varphi_b \\ \varphi_c \end{bmatrix}. \tag{95}$$

This, too, can have an inverse, the derivation of which requires the third equation, for which we again use Sum-Zero.

$$\begin{bmatrix} \varphi_d \\ \varphi_q \\ 0 \end{bmatrix} = \begin{bmatrix} \cos(-\delta) & \cos(\gamma-\delta) & \cos(-\gamma-\delta) \\ \sin(-\delta) & \sin(\gamma-\delta) & \sin(-\gamma-\delta) \\ 1 & 1 & 1 \end{bmatrix} \cdot \begin{bmatrix} \varphi_a \\ \varphi_b \\ \varphi_c \end{bmatrix} = T_P \cdot \begin{bmatrix} \varphi_a \\ \varphi_b \\ \varphi_c \end{bmatrix}. \tag{96}$$

T_P, above, is the transformation produced by R. H. Park[5], which combines the polar reference change with the Clarke representation. It is also called the *dq0* transformation. The inverse is found to be:

$$T_P^{-1} = \frac{2}{3} \begin{bmatrix} \cos(-\delta) & \sin(-\delta) & 1 \\ \cos(\gamma-\delta) & \sin(\gamma-\delta) & 1 \\ \cos(-\gamma-\delta) & \sin(-\gamma-\delta) & 1 \end{bmatrix}. \tag{97}$$

In the literature, the signs of the matrix elements may be different. This results either from the selection of the direction of α (moving CCW or CW with respect to the *d* axis) or of the position of the *q* axis (either leading or lagging the *d* axis). It's important to know the specifics of the model, and to be sure the math matches the description you have in mind.

Let's stop and consider for a moment that we have come, in our derivation, from Phasors to this *dq* representation of the vectors that describe the rotating flux field (or the slightly more abstract rotating voltage and current vectors). These two methods are similar, in that they are both stationary representations of the activities at hand, but we must not confuse them. $\vec{\varphi}_s{'}$ in the *dq* frame of Figure 17 looks like a Phasor in that it has a magnitude that is set and an angle that is fixed with respect to its co-ordinate axes. $\vec{\varphi}_s{'}$ represents a vector such as a field vector, and its position within a motor frame. The Phasors represent only the magnitudes and electrical phase angles of the parameters.

Transformations

Fortescue

In working with unbalanced systems we find that there is a method with which interesting computations can be done after resolving the unbalanced systems into balanced, symmetrical components. C. L. Fortescue produced a method for doing this decomposition with Phasors in 1918, while studying faults in three-phase equipment. [3] Here is a development of his process that puts it into our own terms, to allow us to see its relationship to our work up to now.

In a poly-phase machine, the Phasors representing the flux due to each of the machine's windings are represented as a coplanar set of directed lines. Each of these Phasors can be represented either as a length and an angular position or as an x and y component, so for a three-phase system there are a total of six variables that are describing the three Phasors.[15] These Phasor variables can describe a system with any steady state unbalance due to either magnitude or phase angle deviations from the balanced case.

In the Fortescue representation, for each of the three (or N) Phasors there are three sub-components. Look ahead to (98) for a tabulation that may help visualize this introduction. For each Phasor of V_{abc}, one component is one of a balanced set of Phasors (V_{a1} V_{b1} V_{c1}), collectively called the *positive sequence* set. A second component is one of another balanced set of Phasors (V_{a2} V_{b2} V_{c2}), collectively called the *negative sequence* set. A third (V_{a0} V_{b0} V_{c0}), is one of a set of *identical* Phasors, called the *zero sequence* set. In (98), the first column contains the zero sequence terms, the second and third columns have the positive and negative sequence sets, respectively. (We use this arrangement to agree with the description in the Fortescue paper.)

The positive sequence Phasors are a balanced set describing signals with the same phase sequence as the original unbalanced set. The negative sequence Phasors are also balanced and with their own magnitude, but with the opposite phase sequence. The zero sequence Phasors are identical components applied to each phase — they have no sequence. Each phase of the unbalanced system can therefore be described by the sum of its respective positive-, negative- and zero-sequence components. We will discuss this in more detail as we proceed.

The Fortescue Definition

First let's try to summarize the Fortescue work. He seems to have made an observation during his studies of unbalanced systems, then produced a mathematical demonstration of what he found. Here we have the Phasor definition with which he began, in which he represents three (or N, as it works for any number) Phasors with one set of three components for each phase, for a total of $3N$ components (nine in the three-phase case):

---

[15] *Again, this will also work for more than three phases a, b, c, d, e..., with positive, negative and zero-sequence components for each phase.*

Transformations

$$\underline{V}_a = \underline{V}_{a_0} + \underline{V}_{a_1} + \underline{V}_{a_2}$$
$$\underline{V}_b = \underline{V}_{b_0} + \underline{V}_{b_1} + \underline{V}_{b_2} \qquad (98)$$
$$\underline{V}_c = \underline{V}_{c_0} + \underline{V}_{c_1} + \underline{V}_{c_2}$$
$$\vdots \quad \vdots \quad \vdots \quad \vdots$$

Here we are using voltages, but the process applies to other variables as well. For a three phase system::

- $\underline{V}_{abc}$ are the original Phasors, which we could plot as a polar array in a Phasor diagram..
- $\underline{V}_{a_0 b_0 c_0}$, the first column of (98), are the *zero-sequence* set — these are *defined* to be identical in each phase;
- $\underline{V}_{a_1 b_1 c_1}$ in the second column, are the *positive-sequence* set — this is a balanced set of Phasors with angular offsets (γ) between phases in the same sequence as the originals, $\underline{V}_{abc}$;
- $\underline{V}_{a_2 b_2 c_2}$ in the third column, are the *negative-sequence* set — another balanced set with phase offsets producing the sequence opposite to the originals.

So, C. L. Fortescue *asserts* that an unbalanced set of N Phasors can be represented by two balanced sets of N Phasors plus another set of N Phasors that are the same in all N of the original Phasors, as shown in (98) for $N = 3$. In the following demonstration we will see how the definitions above lead to a useful transformation and its inverse.

A reminder: The upcoming derivation applies to Phasors, not Phase-variables. This will provide the CLF description given in his paper. We will follow the Phasor development with the derivation of a description of unbalanced systems using our exponential rotators, that fits into the context of our study and prepares us for a Phase-variable development of CLF, with which we conclude. The Phase-variable study allows more insight into the issue of dynamic unbalance.

The Phasor Demonstration

We re-write C. L. Fortescue's definition here, to keep it handy:

$$\underline{V}_a = \underline{V}_{a_0} + \underline{V}_{a_1} + \underline{V}_{a_2}$$
$$\underline{V}_b = \underline{V}_{b_0} + \underline{V}_{b_1} + \underline{V}_{b_2} \qquad (99)$$
$$\underline{V}_c = \underline{V}_{c_0} + \underline{V}_{c_1} + \underline{V}_{c_2}$$

First, consider the zero-sequence set: The sum of the original, unbalanced Phasors, $\underline{V}_a + \underline{V}_b + \underline{V}_c$, is what we will call the common voltage – it will be zero for a balanced

Transformations

set of Phasors, but in general there will be a non-zero result, which will have its own phase angle. Since the sums of each of the positive and negative-sequence voltage sets will always be zero — they are balanced by definition — it follows that the sum of just the zero-sequence voltages must also equal the common voltage. For each of the three (or N) phases to have the very same zero-sequence, then, the zero-sequence term in each leg must be 1/3 (or $1/N$) of the common voltage, all with the same phase angle.

Now look at the other two sets of components. They are each a balanced set with its own respective magnitude; and in both, the phase separation is our $\gamma = 2\pi/3$, or $2\pi/N$.

The a, b and c components of each column can therefore be written in terms of each other for simplification. We will arbitrarily choose $\underline{V}_a$ as the reference Phasor and again write the CLF components, this time applying a set of our rotators to the terms to produce all the rows using only the $\underline{V}_a$ sub-components:

$$\underline{V}_a = e^{j0}\underline{V}_{a_0} + e^{j0}\underline{V}_{a_1} + e^{j0}\underline{V}_{a_2}$$
$$\underline{V}_b = e^{j0}\underline{V}_{a_0} + e^{-j\gamma}\underline{V}_{a_1} + e^{j\gamma}\underline{V}_{a_2} \qquad (100)$$
$$\underline{V}_c = e^{j0}\underline{V}_{a_0} + e^{j\gamma}\underline{V}_{a_1} + e^{-j\gamma}\underline{V}_{a_2}.$$

We have again used $-\gamma$ as 2γ and e^{j0} as the unit zero-angle rotator. The $\underline{V}_{abc}$ equations are now written in terms of only $\underline{V}_{a0}$, $\underline{V}_{a1}$, $\underline{V}_{a2}$ and the rotation coefficients.

To match the CLF description in his paper, we can again re-write these Phasors, omitting the under bars and using the definition of a we produced on page 8:

$$V_a = V_{a_0} + V_{a_1} + V_{a_2}$$
$$V_b = V_{a_0} + a^2 V_{a_1} + a V_{a_2} \qquad (101)$$
$$V_c = V_{a_0} + a V_{a_1} + a^2 V_{a_2}.$$

Now put these equations into matrix notation for convenience:

$$\begin{bmatrix} V_a \\ V_b \\ V_c \end{bmatrix} \stackrel{\text{def}}{=} \begin{bmatrix} 1 & 1 & 1 \\ 1 & a^2 & a \\ 1 & a & a^2 \end{bmatrix} \cdot \begin{bmatrix} V_{a_0} \\ V_{a_1} \\ V_{a_2} \end{bmatrix} = (CLF^{-1}) \cdot \begin{bmatrix} V_{a_0} \\ V_{a_1} \\ V_{a_2} \end{bmatrix}. \qquad (102)$$

Also seen in other writings as:

Transformations

$$(CLF^{-1}) \cdot \begin{bmatrix} V_0 \\ V_1 \\ V_2 \end{bmatrix} \quad \text{or} \quad (CLF^{-1}) \cdot \begin{bmatrix} V_0 \\ V_+ \\ V_- \end{bmatrix}. \tag{103}$$

This says that if we are correct in our presumption that we already have the CLF symmetrical components, we can apply an operation to them that will produce the original unbalanced Phasors from which they came. We will therefore call this 3×3 (or 3×N) matrix the CLF $^{-1}$. Now, the CLF $^{-1}$ can itself be inverted to produce the desired CLF process. With the CLF we can produce his symmetrical components, V_{012}, from the inputs, V_{abc}. We have, then, backed into a definition of the Fortescue decomposition and its inverse. Here is the relationship between CLF and its inverse:

$$CLF^{-1} = \begin{bmatrix} 1 & 1 & 1 \\ 1 & a^2 & a \\ 1 & a & a^2 \end{bmatrix}, \text{ so}$$

$$CLF = \begin{bmatrix} 1 & 1 & 1 \\ 1 & a^2 & a \\ 1 & a & a^2 \end{bmatrix}^{-1} = \frac{1}{3} \begin{bmatrix} 1 & 1 & 1 \\ 1 & a & a^2 \\ 1 & a^2 & a \end{bmatrix}, \tag{104}$$

which can be confirmed by performing the multiplication CLF·CLF $^{-1}$, while keeping in mind that $a^3 = a^0 = 1$, and $a^4 = a$. (Use Figure 18, in which the terms are applying direction to unit Phasors, to help in the process.)

(… or, of course, you can do the Gaussian elimination of the augmented matrix to find the inverse directly!)

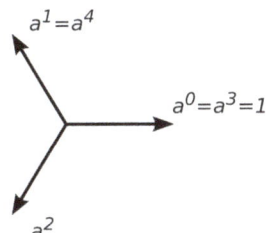

Figure 18: CLF rotators

… but what have we done, here?

We have done a back-door derivation of a method for determining balanced components of a set of Phasors after assuming they exist, and then applied the conversion to its inverse to convince ourselves that it works mathematically. What, though, is actually going on? We can provide an intuitive look at the process with the help of our exponential rotator, $a = e^{j\gamma}$. We will

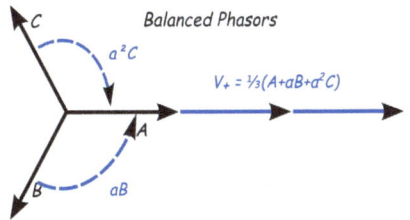

Figure 19: Pos Seq = A+B+C

take a first look using a balanced set of Phasors with a positive phase sequence (black Phasors in Figure 19). If we choose one phase (A) as a reference, and apply the positive rotation to B, then the negative to C, all three Phasors will then be

Transformations

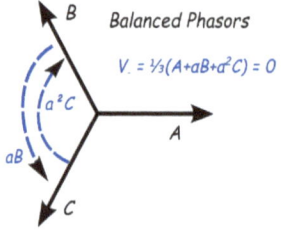

Figure 20: Neg Seq = 0

aligned in the direction of A. Then if we add the three, the result for this example is 3 times the magnitude and still in the direction of A. This result, when divided by three, is in fact a per-phase value such as we defined on page 14. If we do the same process, but with a negative sequence for the black Phasors (Figure 20), B and C will just swap positions, and the sum of the resulting three balanced Phasors is still zero. For any balanced set of positive-sequence Phasors, applying our rotators this way will produce a sum for V_+, while applying them to the negative sequence Phasors will produce zero for V_-. If we had started with a *balanced negative* sequence set, however, this same process would have extracted a sum for V_-, while producing zero for V_+.

Now for the zero-sequence Phasor set (Figure 21): These have the same direction and magnitude for each phase. When using the CLF matrix, we apply *no* rotation in time to the zero-sequence components, and the result is that the same component is applied in each phase.

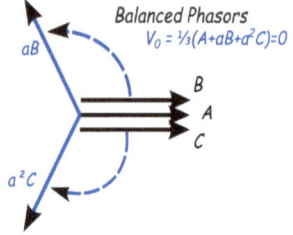

Figure 21: Zero Seq = 0

These rotations and sums are just what the CLF process is doing when it is applied. In an unbalanced system, positive and negative sequence components each add to a resultant after the rotations; and the common voltage is produced with no sequential rotation applied. Division by three returns the correct magnitudes for the component parts. The magnitudes of the three Phasors that are the result of the CLF process are the average of their respective component values.

In our first example, only the positive sequence (+) Phasor exists. An unbalanced Phasor system, for another example, will have errors in phase or magnitude in one or more phases. For Sum-Zero systems with an error, the V_0 will still be zero, although V_- and V_+ will exist. For a Sum-Zero set, $\underline{V}_a$ and $\underline{V}_c$ will both have to change along with $\underline{V}_b$ such that the sum of the Phasors would always be zero.

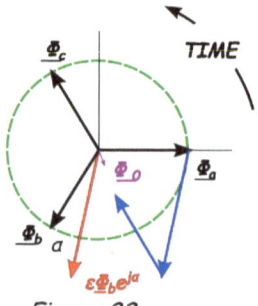

Figure 22: Zero Sequence

Let's look at a system with errors, say in Φ_b. It's not hard to see in Figures 22-24 that this is not Sum-Zero (use the red Φ_b with the errors). Φ_0 (the small Phasor in Figure 22) will be in the direction shown, with magnitude one third of the sum of the three Phasors. Φ_+ in Figure 23 and Φ_- in Figure 24 will be the Phasor averages determined after their respective rotations. In these Figures the blue lines show the way the Phasors add after the rotations. Each of these per-phase resultants, because of the differences in magnitude and phase angle, will behave differently in an equation involving, for example, an impedance.

Transformations

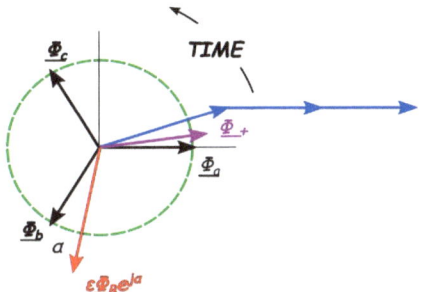
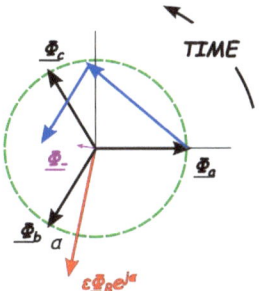

Figure 23: Pos Sequence

Figure 24: Neg Sequence

This has been a pretty nearly complete examination of the determination of CLF with Phasors. Next we will look at CLF for Phase-variables. We will soon see how the counter-rotating line segments forming the ellipse we derived on pp. 17 & 18 are related to the Fortescue positive and negative sequence terms, respectively. We will also discuss zero sequence contributions that are produced by unequal magnitudes or angles of the phase parameters.

The Phase-variable Demonstration

To complete our CLF demonstration we will do another derivation using Phase-variables rather than Phasors. We like Phase-variables because they include the dynamic information of the system, and are signals we can observe and manage. Phase-variables are not limited to single-sine waveforms as are Phasors. Phase-variables can represent sums of sinusoids of different frequencies, and therefore also products of sinusoids. (Remember sine products produce sum and difference frequency components that are themselves summed.) This in turn enables Phase-variables to work for any shape signal, since all signals can be described as Fourier sums of single sinusoids!

We will use Φ as our variable, since flux is an important parameter for the Phase-vectors. Keep in mind that a Phase-variable, with which we can work using instrumentation ('scope, amplifier...), is like a Phase-vector with a zero angle in the spatial rotator. We want to show that any unbalanced set of these pseudo Phase-vectors can be represented by conjugate sets of pairs of counter-rotating, balanced Phase-vectors. This we will do by examining a set of Phase-variables that are balanced, then introducing an unbalance by changing one of them. Then we can place the Phase-variables in space (producing Phase-vectors) and look at the locus of the resultant Space-vector. We begin by recalling the description we developed between Phasors and Phase-Variables at (20) on page 12. (Also see (105) below.)

We can apply a modified CLF to these Phase-variables with a method we will devise based on the CLF for the Phasors. Although these Phase-variables have no directional assignment in *space*, we can assign a direction (say, along the real axis) to them in a *time* plane, which puts them in the form of zero-angle Phase-vectors. OK – let's devise a method for the CLF-style decomposition of these guys. These new

Transformations

Phase-vectors are now aligned with the horizontal axis (but will still add to zero if they are balanced), so it may seem at first glance that there is not much way to do CLF on them — or is there? If we consider the conjugate components separately, CLF begins to make some sense. We devised a way to do this on page 12. Write them this way:

$$\varphi_a = \Phi_a \cos \omega t = \tfrac{1}{2}[\underline{\Phi}_a e^{j\omega t} + \underline{\Phi}_a^* e^{-j\omega t}]$$
$$\varphi_b = \Phi_b \cos(\omega t - \gamma) = \tfrac{1}{2}[\underline{\Phi}_b e^{j\omega t} + \underline{\Phi}_b^* e^{-j\omega t}] \quad (105)$$
$$\varphi_c = \Phi_c \cos(\omega t + \gamma) = \tfrac{1}{2}[\underline{\Phi}_c e^{j\omega t} + \underline{\Phi}_c^* e^{-j\omega t}].$$

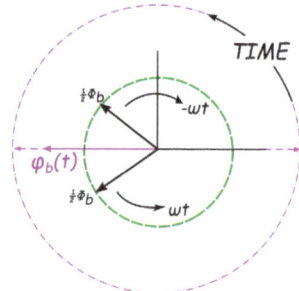

Figure 25: Vector Components of a time-varying Phase-variable

In this description, the conjugate Phasors are now being rotated in time to produce a description of the Phase-variables. When the equations are written this way, the time based extractors will work as they did in the Phasor development above. Figure 25 is an illustration of this scheme. This enables us to devise a method for Phase-variables in which we apply CLF to the CCW parts and CLF* to the conjugate CW components of the cosines, respectively. This is amounts to forcing a change of the electrical phase angle of the Phase-variable (or rather, zero-angle Phase-vector) in the time domain, because the conjugate parts are being moved in opposite directions.

This process produces each of the per-phase values for the positive, negative and zero sequence components. (Remember the per-phase definition beginning on page 14.) To simplify the derivation, we will again start with an example using a positive sequence (CCW) balanced input (with $\Phi_a = \Phi_b = \Phi_c = \Phi$); so in this first example only φ_+ will be non-zero.

We begin to use the {} operator in (106) to mean that in our transformation definition the sign of the exponential multiplier is opposite for the conjugates in this *time* distribution for the Phase-variables. On the next page we start with the nine (eighteen, really, because of the conjugates) components making up three Phase-variables, written in exponential form with the {CLF} multipliers written using the Fortescue form (with $a = e^{j\gamma}$):

Transformations

$$\begin{bmatrix} \varphi_0 \\ \varphi_+ \\ \varphi_- \end{bmatrix} = \{CLF\} \begin{bmatrix} \varphi_a \\ \varphi_b \\ \varphi_c \end{bmatrix} = \frac{1}{3} \begin{vmatrix} \frac{1}{2}\Phi_a\left(e^{j0}e^{j\omega t}+e^{-j0}e^{-j\omega t}\right) \\ +\frac{1}{2}\Phi_b\left(e^{-j\gamma}e^{j\omega t}+e^{j\gamma}e^{-j\omega t}\right) \\ +\frac{1}{2}\Phi_c\left(e^{j\gamma}e^{j\omega t}+e^{-j\gamma}e^{-j\omega t}\right) \\ \\ \frac{1}{2}\Phi_a\left(e^{j0}e^{j\omega t}+e^{-j0}e^{-j\omega t}\right) \\ +\frac{1}{2}\Phi_b\left(ae^{-j\gamma}e^{j\omega t}+a^2 e^{j\gamma}e^{-j\omega t}\right) \\ +\frac{1}{2}\Phi_c\left(a^2 e^{j\gamma}e^{j\omega t}+ae^{-j\gamma}e^{-j\omega t}\right) \\ \\ \frac{1}{2}\Phi_a\left(e^{j0}e^{j\omega t}+e^{-j0}e^{-j\omega t}\right) \\ +\frac{1}{2}\Phi_b\left(a^2 e^{-j\gamma}e^{j\omega t}+ae^{j\gamma}e^{-j\omega t}\right) \\ +\frac{1}{2}\Phi_c\left(ae^{j\gamma}e^{j\omega t}+a^2 e^{-j\gamma}e^{-j\omega t}\right) \end{vmatrix}, \quad (106)$$

so that after bringing the ½ outside the brackets and completing the exponential products, we have, for the balanced case:

$$\begin{bmatrix} \varphi_0 \\ \varphi_+ \\ \varphi_- \end{bmatrix} = \frac{1}{6} \begin{vmatrix} \Phi_a\left(e^{j\omega t}+e^{-j\omega t}\right) \\ +\Phi_b\left(e^{-j\gamma}e^{j\omega t}+e^{j\gamma}e^{-j\omega t}\right) \\ +\Phi_c\left(e^{j\gamma}e^{j\omega t}+e^{-j\gamma}e^{-j\omega t}\right) \\ \\ \Phi_a\left(e^{j\omega t}+e^{-j\omega t}\right) \\ +\Phi_b\left(e^{j\omega t}+e^{-j\omega t}\right) \\ +\Phi_c\left(e^{j\omega t}+e^{-j\omega t}\right) \\ \\ \Phi_a\left(e^{j0}e^{j\omega t}+e^{-j0}e^{-j\omega t}\right) \\ +\Phi_b\left(e^{j\gamma}e^{j\omega t}+e^{-j\gamma}e^{-j\omega t}\right) \\ +\Phi_c\left(e^{-j\gamma}e^{j\omega t}+e^{j\gamma}e^{-j\omega t}\right) \end{vmatrix} = \frac{1}{3} \begin{bmatrix} 0 \\ 3\Phi \cdot \cos\omega t \\ 0 \end{bmatrix}. \quad (107)$$

Now we can see the way all the components add, for this balanced case, to the per-phase Phase-variable for the positive sequence, since in that case $\Phi_a = \Phi_b = \Phi_c = \Phi$. For the unbalanced case, though, the process will generally produce all three φ_{0+-} variables at this point. These will be the values for the per-phase representation of the {CLF} symmetrical components for Phase-variables; that is the magnitudes of the zero, positive and negative sequence sets of components. As we have seen, for the

Transformations

balanced case only one component set has a non-zero sum. With these per-phase values we can build the symmetrical Phase-vectors that will sum to the same Space-vector as the original unbalanced set.

As with the Phasor CLF, we can express each of the legs of each balanced set in terms of a chosen leg. We choose phase a, to which have assigned no rotation. It is therefore the same mathematically as the per-phase result above in (107). Having found the per-phase values of the φ_{0+-} Phase-variables, the conjugate rotators now make the change in time (or electrical phase angle), and the spatial rotators place these new Phase-vectors on the spatial plane, using the γ_s distribution of the machine windings. Once this is done we can do the vector sum of the new Phase-vectors to arrive at the same Space-vector that will be produced by the original unbalanced set. The same spatial distribution is applied to each variable of the CLF component set, because it is effected by the placement of the machine windings.

Here are some things to note:

> φ_o, having the same size and direction for each phase, will always have a zero resultant when given a spatial γ distribution among the windings.
>
> φ_+, after our γ distribution, will be balanced, and its vector sum will produce a Space-vector component with a CCW circular locus.
>
> φ_-, if it is not zero, will be arrayed similarly and balanced, but with the opposite phase sequence. The resulting Space-vector would in that case have a component with a CW circular locus.

Remember, DO NOT CONFUSE THE TIME- AND SPACE-DOMAIN PLOTS! We have done the {CLF} rotations in *time* to find the per-phase values, operating on the Phase-variables. The positive and negative sequence per-phase values are then given new electrical phase angles, creating balanced sets. Then we arrange each Phase-variable in *space* to produce the Phase-vectors. Let us next look at the process for the unbalanced case in some detail, to help with our intuitive development. We start with a look at the negative sequence Phase-variable.

From (107), we can write the per-phase sum for just the negative sequence Phase-variable:

$$\varphi_- = \frac{1}{6} \begin{bmatrix} \Phi_a\left(e^{j0}e^{j\omega t}+e^{-j0}e^{-j\omega t}\right) \\ + \Phi_b\left(e^{j\gamma}e^{j\omega t}+e^{-j\gamma}e^{-j\omega t}\right) \\ + \Phi_c\left(e^{-j\gamma}e^{j\omega t}+e^{j\gamma}e^{-j\omega t}\right) \end{bmatrix}. \quad (108)$$

Since this is a per-phase value, we need first to distribute this value in the *time* plane to produce the balanced three negative sequence Phase-variables. Since the decomposition of the negative sequence used {1 a² a} to do the rotation on the original Phase-variables, we must distribute the new per-phase values back to the same location (using {1 a a²}). We have factored the exponential terms as we write:

Transformations

$$\varphi_-^a = \{1\} \; \varphi_- = \frac{1}{6}\left[e^{j\omega t}(\Phi_a + \Phi_b e^{j\gamma} + \Phi_c e^{-j\gamma}) + e^{-j\omega t}(\Phi_a + \Phi_b e^{-j\gamma} + \Phi_c e^{j\gamma})\right]$$

$$\varphi_-^b = \{a\} \; \varphi_- = \frac{1}{6}\left[e^{j\omega t}(\Phi_a e^{j\gamma} + \Phi_b e^{-j\gamma} + \Phi_c) + e^{-j\omega t}(\Phi_a e^{-j\gamma} + \Phi_b e^{j\gamma} + \Phi_c)\right] \quad (109)$$

$$\varphi_-^c = \{a^2\} \; \varphi_- = \frac{1}{6}\left[e^{j\omega t}(\Phi_a e^{-j\gamma} + \Phi_b + \Phi_c e^{j\gamma}) + e^{-j\omega t}(\Phi_a e^{j\gamma} + \Phi_b + \Phi_c e^{-j\gamma})\right]$$

Here again, the braces infer application of the conjugate rotators to the conjugate terms, to effect the rotations in the time plane. After the time rotations, the *spatial* distribution will produce the new, balanced negative sequence Phase-vectors.

First, though, it is interesting to look a bit more at the factored exponential form of (108), which is the negative sequence result of the {CLF} decomposition:

$$\varphi_- = \frac{1}{6}\left[e^{j\omega t}(\Phi_a + \Phi_b e^{j\gamma} + \Phi_c e^{-j\gamma}) + e^{-j\omega t}(\Phi_a + \Phi_b e^{-j\gamma} + \Phi_c e^{j\gamma})\right]. \quad (110)$$

Now by putting it back into trig form we can see that the time sequence in φ_- has been reversed from the input sequence of φ by the {CLF} process:

$$\varphi_- = \frac{1}{3}\left(\Phi_a \cos\omega t + \Phi_b \cos(\omega t + \gamma) + \Phi_c \cos(\omega t - \gamma)\right). \quad (111)$$

Later, when we look at the effects of adding errors to our inputs, we will watch for this change.

Now back to our development. We can already see that the parts of (110) and (111) will each add to zero for the balanced case, but let us continue to expand the *a*, *b* and *c* terms, to look at the three negative sequence Phase-vector components. We continue to use lower case *abc* here to denote the new, still balanced but now *spatially* offset components:

$$\vec{\varphi}_-^a = \overrightarrow{e^{j0}} \; \varphi_-^a = \frac{1}{6}\left[e^{j\omega t}(\Phi_a + \Phi_b e^{j\gamma} + \Phi_c e^{-j\gamma}) + e^{-j\omega t}(\Phi_a + \Phi_b e^{-j\gamma} + \Phi_c e^{j\gamma})\right]$$

$$\vec{\varphi}_-^b = \overrightarrow{e^{j\gamma}} \; \varphi_-^b = \frac{1}{6}\left[e^{j\omega t}(\Phi_a e^{-j\gamma} + \Phi_b + \Phi_c e^{j\gamma}) + e^{-j\omega t}(\Phi_a + \Phi_b e^{-j\gamma} + \Phi_c e^{j\gamma})\right] \quad (112)$$

$$\vec{\varphi}_-^c = \overrightarrow{e^{-j\gamma}} \; \varphi_-^c = \frac{1}{6}\left[e^{j\omega t}(\Phi_a e^{j\gamma} + \Phi_b e^{-j\gamma} + \Phi_c) + e^{-j\omega t}(\Phi_a + \Phi_b e^{-j\gamma} + \Phi_c e^{j\gamma})\right]$$

(Vector arrow over the result omitted.)

Notice that the CCW terms must always add to zero here, because the terms in the columns have equal spatial distribution. The CW terms in each phase add to zero only if $\Phi_a = \Phi_b = \Phi_c = \Phi$. These are the general form of a balanced set of three Phase-vectors that comprise the negative sequence part of the Space-vector representation.

Transformations

Again, if there is an unbalance in the original system, only the sum of the CW negative sequence set described in (112) will have a non-zero value.

Next we add an error in magnitude and angle to see how the errors are propagated.

If we modify φ_b from (105) to produce some errors in magnitude and phase angle, we will be able to see the way the negative sequence set arises. There can be one or many errors in the system, but we'll just look at phase b, here. (There are many ways to introduce errors; these errors are just one example.) We can write a new φ_b this way as our example:

$$\varphi_b = \varepsilon \cdot \Phi_b \cos(\omega t - \gamma + \upsilon) \ . \tag{113}$$

Here we have added a magnitude error factor ε and an angular error υ to phase b. Now we can rewrite φ_- from (111) in trig form to include the errors:

$$\varphi_{-\varepsilon} = \frac{1}{3}\left(\Phi_a \cos\omega t + \varepsilon \cdot \Phi_b \cos(\omega t + \gamma + \upsilon) + \Phi_c \cos(\omega t - \gamma)\right) \ . \tag{114}$$

Watch out for the sign of γ in the b term here. In (113) it is negative because of the original phase sequence in time; while in (114) it is positive after having endured the {CLF} process (remember, $-\gamma -\gamma = \gamma$).

We can eliminate the Φ_a and Φ_c terms in this equation by adding and subtracting the original (error free) φ_b term (in bold), and then doing the sum. *Remember the phase magnitudes (φ_{abc}) are equal here because the errors are explicitly introduced*:

$$\begin{aligned}\varphi_{-\varepsilon} &= \frac{\Phi_-}{3}\left[\cos\omega t + \cos(\omega t + \gamma) + \cos(\omega t - \gamma)\right. \\ &\quad \left. + \varepsilon \cdot \cos(\omega t + \gamma + \upsilon) - \cos(\omega t + \gamma)\right] \\ &= \frac{\Phi_-}{3}\left[\varepsilon \cdot \cos(\omega t + \gamma + \upsilon) - \cos(\omega t + \gamma)\right]\end{aligned} \tag{115}$$

This is the negative sequence per-phase Phase-variable that has arisen because of the errors in phase b. We can rewrite the no longer zero valued φ_- as we factor the exponential form:

$$\varphi_{-\varepsilon} = \frac{\Phi_-}{6}\left[\varepsilon \cdot \left(e^{j(\omega t + \gamma + \upsilon)} + e^{-j(\omega t + \gamma + \upsilon)}\right) - \left(e^{j(\omega t + \gamma)} + e^{-j(\omega t + \gamma)}\right)\right] \ . \tag{116}$$

Now we can do the *time* distribution of this Phase-variable that includes the errors, as we did in (109):

Transformations

$$\varphi_{-\varepsilon}^{a} = \{1\} \; \varphi_{-\varepsilon} = \frac{\Phi_{-}}{6}[\varepsilon \cdot (e^{j(\omega t+\gamma+\upsilon)} + e^{-j(\omega t+\gamma+\upsilon)}) - (e^{j(\omega t+\gamma)} + e^{-j(\omega t+\gamma)})]$$

$$\varphi_{-\varepsilon}^{b} = \{a\} \; \varphi_{-\varepsilon} = \frac{\Phi_{-}}{6}[\varepsilon \cdot (e^{j(\omega t-\gamma+\upsilon)} + e^{-j(\omega t-\gamma+\upsilon)}) - (e^{j(\omega t-\gamma)} + e^{-j(\omega t-\gamma)})] \; . \quad (117)$$

$$\varphi_{-\varepsilon}^{c} = \{a^2\} \; \varphi_{-\varepsilon} = \frac{\Phi_{-}}{6}[\varepsilon \cdot (e^{j(\omega t+\upsilon)} + e^{-j(\omega t+\upsilon)}) - (e^{j(\omega t)} + e^{-j(\omega t)})]$$

Don't forget that the {} operator means the sign of the exponential multiplier is opposite for the conjugates in the time distribution for the Phase-variables.

Finally, we can add the *spatial* distribution to build the Phase-vectors for the negative sequence:

$$\overrightarrow{\varphi_{-\varepsilon}^{a}} = \overrightarrow{e^{j0}} \varphi_{-\varepsilon}^{a} = \frac{\Phi_{-}}{6}[\varepsilon \cdot (e^{j(\omega t+\gamma+\upsilon)} + e^{-j(\omega t+\gamma+\upsilon)}) - (e^{j(\omega t+\gamma)} + e^{-j(\omega t+\gamma)})]$$

$$\overrightarrow{\varphi_{-\varepsilon}^{b}} = \overrightarrow{e^{j\gamma}} \varphi_{-\varepsilon}^{b} = \frac{\Phi_{-}}{6}[\varepsilon \cdot (e^{j(\omega t+\upsilon)} + e^{-j(\omega t+\gamma+\upsilon)}) - (e^{j(\omega t)} + e^{-j(\omega t+\gamma)})] \quad . \quad (118)$$

$$\overrightarrow{\varphi_{-\varepsilon}^{c}} = \overrightarrow{e^{j-\gamma}} \varphi_{-\varepsilon}^{c} = \frac{\Phi_{-}}{6}[\varepsilon \cdot (e^{j(\omega t-\gamma+\upsilon)} + e^{-j(\omega t+\gamma+\upsilon)}) - (e^{j(\omega t-\gamma)} + e^{-j(\omega t+\gamma)})]$$

(Vector arrow over the result omitted.)

When we add these components and factor, we have at last:

$$\overrightarrow{\varphi_{-\varepsilon}} = \overrightarrow{\varphi_{-\varepsilon}^{a}} + \overrightarrow{\varphi_{-\varepsilon}^{b}} + \overrightarrow{\varphi_{-\varepsilon}^{c}} = \frac{\Phi_{-}}{2}[\overrightarrow{e^{-j(\omega t+\gamma)}}(\varepsilon \cdot e^{-j\upsilon} - 1)] \; . \quad (119)$$

This is our desired new Phase-vector due to the negative sequence set, rotating in the clockwise (negative) direction. Checking, we see that if there is no magnitude error ($\varepsilon = 1$) and no phase error ($\upsilon = 0$), the negative rotating Space-vector disappears as it should.

This is the non-zero right hand (CW) component we see in (43), back in the discussion of Space-vectors.

Next let's do the same process for the positive sequence, and examine the result. From (107), we can write the sum for the per-phase positive sequence, as we did for the negative:

$$\varphi_{+} = \frac{1}{6}\left[\Phi_a(e^{j\omega t} + e^{-j\omega t}) + \Phi_b(e^{j\omega t} + e^{-j\omega t}) + \Phi_c(e^{j\omega t} + e^{-j\omega t})\right] \; . \quad (120)$$

We will again make the *temporal* distribution to produce the new balanced positive sequence Phase-variables. First we factor the exponential form:

$$\varphi_{+} = \frac{1}{6}\left[e^{j\omega t}(\Phi_a + \Phi_b + \Phi_c) + e^{-j\omega t}(\Phi_a + \Phi_b + \Phi_c)\right] \; . \quad (121)$$

Here we can see that the conjugate parts of this equation do *not* add to zero for the

Transformations

balanced case, as they did for the negative sequence.

Let us again carry on and find the three positive sequence components, to see how they look. For the rotations in time, we again use the inverse of the {CLF} rotations:

$$\varphi_+^a = \{1\}\varphi_+ = \frac{1}{6}\left[e^{j\omega t}(\Phi_a+\Phi_b+\Phi_c)+e^{-j\omega t}(\Phi_a+\Phi_b+\Phi_c)\right]$$

$$\varphi_+^b = \{a^2\}\varphi_+ = \frac{1}{6}\left[e^{j\omega t}(\Phi_a e^{-j\gamma}+\Phi_b e^{-j\gamma}+\Phi_c e^{-j\gamma})+e^{-j\omega t}(\Phi_a e^{-j\gamma}+\Phi_b e^{-j\gamma}+\Phi_c e^{-j\gamma})\right] \quad (122)$$

$$\varphi_+^c = \{a\}\varphi_+ = \frac{1}{6}\left[e^{j\omega t}(\Phi_a e^{j\gamma}+\Phi_b e^{j\gamma}+\Phi_c e^{j\gamma})+e^{-j\omega t}(\Phi_a e^{j\gamma}+\Phi_b e^{j\gamma}+\Phi_c e^{j\gamma})\right].$$

We again use lower case *abc* to denote the new, balanced, spatially offset Phase-vectors:

$$\vec{\varphi}_+^a = e^{j0}\varphi_+^a = \frac{1}{6}\left[e^{j\omega t}(\Phi_a+\Phi_b+\Phi_c)+e^{-j\omega t}(\Phi_a+\Phi_b+\Phi_c)\right]$$

$$\vec{\varphi}_+^b = e^{j\gamma}\varphi_+^b = \frac{1}{6}\left[e^{j\omega t}(\Phi_a+\Phi_b+\Phi_c)+e^{-j\omega t}(\Phi_a e^{-j\gamma}+\Phi_b e^{-j\gamma}+\Phi_c e^{-j\gamma})\right] \quad (123)$$

$$\vec{\varphi}_+^c = e^{-j\gamma}\varphi_+^c = \frac{1}{6}\left[e^{j\omega t}(\Phi_a+\Phi_b+\Phi_c)+e^{-j\omega t}(\Phi_a e^{j\gamma}+\Phi_b e^{j\gamma}+\Phi_c e^{j\gamma})\right].$$

(Vector arrow over the result omitted.)

For the positive sequence, the CW terms always sum to zero, because the terms in the columns are evenly distributed. When $\Phi_a=\Phi_b=\Phi_c=\Phi$, the vector sum must be $3/2\ \Phi e^{j\omega t}$. These are the general form of a set of three Phase-vectors that comprise the positive sequence components of the Space-vector representation. If there is an unbalance in the original system, the sum of the CCW positive sequence set described in (123) will have a different value. The process will return a different solution for the CCW positive sequence set including an error, although the CW terms will still add to zero.

If we again use φ_b with the same errors in magnitude and phase angle as before, we will be able to see the way the errors modify the positive sequence set. So again let:

$$\varphi_b = \varepsilon\cdot\Phi_b\cos(\omega t-\gamma+\upsilon). \quad (124)$$

Now we can rewrite φ_+ from (107) in trig form to include the errors:

$$\varphi_+ = \frac{1}{3}\left(\Phi_a\cos(\omega t)+\varepsilon\cdot\Phi_b\cos(\omega t+\upsilon) + \Phi_c\cos(\omega t)\right). \quad (125)$$

Notice there is no γ in the *b* term here. In (124) it is negative because of the original phase sequence in time; while in (125) it is gone after having endured the {CLF} process (this time use $-\gamma+\gamma = 0$). We can simplify this equation by adding and subtracting the original (error free) φ_b term, and doing the sum. Remember the phase magnitudes are equal here, too:

Transformations

$$\varphi_+ = \frac{\Phi_+}{3}\Big[\cos(\omega t) + \cos(\omega t) + \cos(\omega t)$$
$$+ \varepsilon \cos(\omega t + v) - \cos(\omega t)\Big] \qquad (126)$$
$$= \Phi_+\Big[\cos(\omega t) + \frac{1}{3}(\varepsilon \cdot \cos(\omega t + v) - \cos(\omega t))\Big].$$

For the positive sequence, the sum of the balanced terms produce the per-phase Phase-variable seen in the balanced case, augmented by the error we added. Now, we have a new φ_+ for the positive sequence as we factor the exponential form:

$$\varphi_+ = \frac{\Phi_+}{2}\Big[(e^{j\omega t} + e^{-j\omega t}) + \frac{1}{3}\big(\varepsilon \cdot (e^{j(\omega t + v)} + e^{-j(\omega t + v)}) - (e^{j\omega t} + e^{-j\omega t})\big)\Big]. \qquad (127)$$

As before, we can do the *time* distribution of this Phase-variable:

$$\varphi_+^a = \{1\} \quad \varphi_+ = \frac{\Phi_+}{2}\Big[(e^{j\omega t} + e^{-j\omega t}) \quad + \frac{1}{3}\big(\varepsilon \cdot (e^{j(\omega t + v)} + e^{-j(\omega t + v)}) - (e^{j\omega t} + e^{-j\omega t})\big)\Big]$$

$$\varphi_+^b = \{a^2\} \quad \varphi_+ = \frac{\Phi_+}{2}\Big[(e^{j(\omega t - \gamma)} + e^{-j(\omega t - \gamma)}) + \frac{1}{3}\big(\varepsilon \cdot (e^{j(\omega t - \gamma + v)} + e^{-j(\omega t - \gamma + v)}) - (e^{j(\omega t - \gamma)} + e^{-j(\omega t - \gamma)})\big)\Big] \qquad (128)$$

$$\varphi_+^c = \{a\} \quad \varphi_+ = \frac{\Phi_+}{2}\Big[(e^{j(\omega t + \gamma)} + e^{-j(\omega t + \gamma)}) + \frac{1}{3}\big(\varepsilon \cdot (e^{j(\omega t + \gamma + v)} + e^{-j(\omega t + \gamma + v)}) - (e^{j(\omega t + \gamma)} + e^{-j(\omega t + \gamma)})\big)\Big].$$

Don't forget: The sequence for the time displacement rotators is opposite those for the negative sequence; and the {} operator means the sign of the exponential multiplier is opposite for the conjugates, in the time distribution for the Phase-variables.

Finally, we can add the *spatial* distribution to build the Phase-vectors for the positive sequence:

$$\vec{\varphi}_+^a = e^{j0}\varphi_+^a = \frac{\Phi_+}{2}\Big[(e^{j\omega t} + e^{-j\omega t}) \quad + \frac{1}{3}\big(\varepsilon \cdot (e^{j(\omega t + v)} + e^{-j(\omega t + v)}) - (e^{j\omega t} + e^{-j\omega t})\big)\Big]$$

$$\vec{\varphi}_+^b = e^{j\gamma}\varphi_+^b = \frac{\Phi_+}{2}\Big[(e^{j\omega t} + e^{-j(\omega t + \gamma)}) + \frac{1}{3}\big(\varepsilon \cdot (e^{j(\omega t + v)} + e^{-j(\omega t + \gamma + v)}) - (e^{j\omega t} + e^{-j(\omega t + \gamma)})\big)\Big] \qquad (129)$$

$$\vec{\varphi}_+^c = e^{-j\gamma}\varphi_+^c = \frac{\Phi_+}{2}\Big[(e^{j\omega t} + e^{-j(\omega t - \gamma)}) + \frac{1}{3}\big(\varepsilon \cdot (e^{j(\omega t + v)} + e^{-j(\omega t - \gamma + v)}) - (e^{j\omega t} + e^{-j(\omega t - \gamma)})\big)\Big].$$

(Vector arrow over the result omitted.)

The spatial rotator sequence is the same for the positive, negative and zero sequence elements since this distribution represents the physical placement of the windings in a machine. When we add these components and factor, we have at last:

$$\vec{\varphi}_+ = \vec{\varphi}_+^a + \vec{\varphi}_+^b + \vec{\varphi}_+^c = e^{j\omega t}\Big(\frac{3}{2}\Phi_+ + \varepsilon \cdot e^{j(v)} - 1\Big), \qquad (130)$$

our erroneous Space-vector, rotating in the counter-clockwise (positive) direction. Checking, we see that if there is no magnitude error ($\varepsilon = 1$) and no phase error ($v =$

55

Transformations

0); then $\vec{\varphi}_S = \vec{\varphi}_+ = \frac{3}{2}\Phi\overrightarrow{e^{j\omega t}}$, as it should be for a balanced system.

This is the left hand (CCW) component we see in (43), back in the discussion of Space-vectors.

These positive and negative sequence components precisely match the description derived at (43), yet we still have the zero sequence components to extract. Let's repeat the process once more for the zero sequence, and resolve this apparent incongruity. Again from (107), this time we write the sum for the zero sequence:

$$\varphi_0 = \frac{1}{6}\left| \begin{array}{c} \Phi_a\left(e^{j\omega t}+e^{-j\omega t}\right) \\ +\Phi_b\left(e^{-j\gamma}e^{j\omega t}+e^{j\gamma}e^{-j\omega t}\right) \\ +\Phi_c\left(e^{j\gamma}e^{j\omega t}+e^{-j\gamma}e^{-j\omega t}\right) \end{array} \right|. \quad (131)$$

There is *no* time distribution for the zero-sequence Phase-variables, because they are exactly the same in all three-phases. We will still factor the exponential form:

$$\varphi_0 = \frac{1}{6}\left[e^{j\omega t}\left(\Phi_a+\Phi_b e^{-j\gamma}+\Phi_c e^{j\gamma}\right)+e^{-j\omega t}\left(\Phi_a+\Phi_b e^{j\gamma}+\Phi_c e^{-j\gamma}\right)\right]. \quad (132)$$

We can again see that the conjugate parts of this equation will each add to zero for the balanced case, but let us carry on and find the three zero sequence components, to see how they compare. We again use the lower case *abc* to denote the new, spatially offset components. These components are the same magnitude, but with our equal spatial separation:

$$\vec{\varphi}_0^a = \frac{1}{6}\left[e^{j\omega t}\left(\Phi_a+\Phi_b e^{-j\gamma}+\Phi_c e^{j\gamma}\right)+e^{-j\omega t}\left(\Phi_a+\Phi_b e^{j\gamma}+\Phi_c e^{-j\gamma}\right)\right]$$

$$\vec{\varphi}_0^b = \frac{1}{6}\left[e^{j\omega t}\left(\Phi_a e^{j\gamma}+\Phi_b+\Phi_c e^{-j\gamma}\right)+e^{-j\omega t}\left(\Phi_a e^{j\gamma}+\Phi_b e^{-j\gamma}+\Phi_C\right)\right]. \quad (133)$$

$$\vec{\varphi}_0^c = \frac{1}{6}\left[e^{j\omega t}\left(\Phi_a e^{-j\gamma}+\Phi_b e^{j\gamma}+\Phi_c\right)+e^{-j\omega t}\left(\Phi_a e^{-j\gamma}+\Phi_b+\Phi_c e^{j\gamma}\right)\right]$$

(Vector arrow over the result omitted.)

Now these terms sum to zero whether the phase (*abc*) coefficients are equal or not, which makes sense because the terms have equal magnitude and distribution.

If we again modify φ_b from (105) to include the same errors in magnitude and phase angle, we will be able to see the way the error propagates. We write the new φ_b the same way:

$$\varphi_b = \varepsilon \cdot \Phi_b \cos(\omega t - \gamma + \upsilon), \quad (134)$$

and have added the same errors to phase *b*. Now we can rewritet φ_0 this way to include the errors:

Transformations

$$\varphi_0 = \frac{1}{3}\left(\Phi_a \cos(\omega t) + \varepsilon \cdot \Phi_b \cos(\omega t - \gamma + \upsilon) + \Phi_c \cos(\omega t + \gamma)\right). \tag{135}$$

Here the signs of γ are unchanged because the {CLF} process did no rotation on this component.

We can again eliminate the Φ_a and Φ_c terms in this equation by adding and subtracting the original (balanced) Φ_b term, and doing the sum — remembering that the phase magnitudes are equal here:

$$\begin{aligned}\varphi_0 &= \frac{\Phi}{3}\Big[\cos\omega t + \cos(\omega t - \gamma) + \cos(\omega t + \gamma) \\ &\quad + \varepsilon \cdot \cos(\omega t - \gamma + \upsilon) - \cos(\omega t - \gamma)\Big] \\ &= \frac{\Phi}{3}\big[\varepsilon \cdot \cos(\omega t - \gamma + \upsilon) - \cos(\omega t - \gamma)\big]\end{aligned} \tag{136}$$

For the zero sequence, the balanced terms add out and only the error remains. Since this is the only part that remains after the summing, we can rewrite the no longer zero valued φ_0 as we factor the exponential form:

$$\varphi_0 = \frac{\Phi}{6}\Big[\varepsilon \cdot \big(e^{j(\omega t - \gamma + \upsilon)} + e^{-j(\omega t - \gamma + \upsilon)}\big) - \big(e^{j(\omega t - \gamma)} + e^{-j(\omega t - \gamma)}\big)\Big]. \tag{137}$$

Again, there is no *time* distribution of this Phase-variable, so we write:

$$\begin{aligned}\varphi_0^{\,a} &= \frac{\Phi}{6}\Big[\varepsilon \cdot \big(e^{j(\omega t - \gamma + \upsilon)} + e^{-j(\omega t - \gamma + \upsilon)}\big) - \big(e^{j(\omega t - \gamma)} + e^{-j(\omega t - \gamma)}\big)\Big] \\ \varphi_0^{\,b} &= \frac{\Phi}{6}\Big[\varepsilon \cdot \big(e^{j(\omega t - \gamma + \upsilon)} + e^{-j(\omega t - \gamma + \upsilon)}\big) - \big(e^{j(\omega t - \gamma)} + e^{-j(\omega t - \gamma)}\big)\Big]. \\ \varphi_0^{\,c} &= \frac{\Phi}{6}\Big[\varepsilon \cdot \big(e^{j(\omega t - \gamma + \upsilon)} + e^{-j(\omega t - \gamma + \upsilon)}\big) - \big(e^{j(\omega t - \gamma)} + e^{-j(\omega t - \gamma)}\big)\Big]\end{aligned} \tag{138}$$

Finally, we can add the *spatial* distribution to build the Phase-vectors for the zero sequence:

$$\begin{aligned}\overrightarrow{\varphi_0^{\,a}} &= e^{j0}\,\varphi_a^{\varepsilon} = \frac{\Phi}{6}\Big[\varepsilon \cdot \big(e^{j(\omega t - \gamma + \upsilon)} + e^{-j(\omega t - \gamma + \upsilon)}\big) - \big(e^{j(\omega t - \gamma)} + e^{-j(\omega t - \gamma)}\big)\Big] \\ \overrightarrow{\varphi_0^{\,b}} &= e^{j\gamma}\,\varphi_b^{\varepsilon} = \frac{\Phi}{6}\Big[\varepsilon \cdot \big(e^{j(\omega t + \upsilon)} + e^{-j(\omega t + \gamma + \upsilon)}\big) - \big(e^{j(\omega t)} + e^{-j(\omega t + \gamma)}\big)\Big] \\ \overrightarrow{\varphi_0^{\,c}} &= e^{-j\gamma}\,\varphi_c^{\varepsilon} = \frac{\Phi}{6}\Big[\varepsilon \cdot \big(e^{j(\omega t + \gamma + \upsilon)} + e^{-j(\omega t + \upsilon)}\big) - \big(e^{j(\omega t + \gamma)} + e^{-j(\omega t)}\big)\Big].\end{aligned} \tag{139}$$

(Vector arrow over the result omitted.)

Transformations

When we add these components and factor, we have at last:

$$\vec{\varphi_0} = \vec{\varphi_0}^a + \vec{\varphi_0}^b + \vec{\varphi_0}^c \equiv 0 , \qquad (140)$$

There is *never* a Space-vector component from the zero-sequence set! This is reasonable, since each component of the zero-sequence set is the same – they must *always* add to zero in space after the spatial distribution. This does not mean that there are no zero sequence currents in the respective phases. There are, and they will cause losses, even though they do not contribute to the flux field.

It is also important to note that the CLF process *presumes* that all the phases do have the very same zero sequence component. We can see that with a voltage error in Phase *b*, there will be additional current components produced. These currents will be distributed according to the connected circuitry, and not necessarily evenly. With $P = I^2R$, the heating in some windings could be substantially different than expected.

Transformations

Conclusion

We have demonstrated a number of the transformations used in Power Systems analyses. These transformations usually require that the system that is being described be balanced, or at least Sum-Zero. This is OK for an open loop waveform generator that controls the amplitude and phase angle with respect to a reference. If used in a feedback waveform correction system to balance poly-phase signals that are not Sum-Zero, the transformations and their inverses will not produce signals that servo amplifiers can use to produce the correction inputs.

The use of transformations applied to Fortescue components of Phase-variables may make it possible to produce correction terms for the three parts of the control system, since the positive and negative-sequence CLF components are always balanced. In fact, our study of CLF shows that by feeding back just the zero sequence components to each phase controller, we can render the system Sum-Zero, thereby enabling the transformations to be useful.

Doing all the derivations within this booklet with a consistent set of math tools should help its readers see the relationships between the several methods, and make it possible to understand their virtues; but their limitations as well.

Bibliography

1] : Clarke, Edith; Circuit analysis of A-C power systems; J. Wiley & sons, inc., 1943
2] : Clarke, Edith; Circuit analysis of A-C power systems; J. Wiley & sons, inc., 1950
3] : Fortescue, C. L., "Method Of Symmetrical Co-Ordinates Applied To The Solution Of Polyphase Networks", American Institute of Electrical Engineers, 1918
4] : Nearing, James; Mathematical Tools for Physics; Dover Publications Inc., 2010
5] : Park, R. H., "Two Reaction Theory of Synchronous Machines", AIEE; 716-730, 1929

www.ingramcontent.com/pod-product-compliance
Lightning Source LLC
Chambersburg PA
CBHW041204180526
45172CB00006B/1193